Interactive
Mathematics Program®

INTEGRATED HIGH SCHOOL MATHEMATICS

Meadows or Malls?

FIRST EDITION AUTHORS:
Dan Fendel, Diane Resek, Lynne Alper, and Sherry Fraser

CONTRIBUTORS TO THE SECOND EDITION:
Sherry Fraser, IMP for the 21st Century
Jean Klanica, IMP for the 21st Century
Brian Lawler, California State University San Marcos
Eric Robinson, Ithaca College, NY
Lew Romagnano, Metropolitan State College of Denver, CO
Rick Marks, Sonoma State University, CA
Dan Brutlag, Meaningful Mathematics
Alan Olds, Colorado Writing Project
Mike Bryant, Santa Maria High School, CA
Jeri P. Philbrick, Oxnard High School, CA
Lori Green, Lincoln High School, CA
Matt Bremer, Berkeley High School, CA
Margaret DeArmond, Kern High School District, CA

Key Curriculum Press

Second Edition   I M P

This material is based upon work supported by the National Science Foundation under award numbers ESI-9255262, ESI-0137805, and ESI-0627821. Any opinions, findings, and conclusions or recommendations expressed in this publication are those of the authors and do not necessarily reflect the views of the National Science Foundation.

Key Curriculum Press
1150 65th Street
Emeryville, California 94608
email: editorial@keypress.com
www.keypress.com
10 9 8 7 6 5 4 3 2 1   14 13 12 11
ISBN 978-1-60440-048-9
Printed in the United States of America

Project Editors
Mali Apple, Josephine Noah, Sharon Taylor

Project Administrators
Emily Reed, Juliana Tringali

Professional Reviewers
Rick Marks, Sonoma State University, CA
D. Michael Bryant, Santa Maria High School, CA, retired

Accuracy Checker
Carrie Gongaware

First Edition Teacher Reviewers
Daniel R. Bennett, Moloka'i High School, HI
Maureen Burkhart, Northridge Academy High School, CA
Dwight Fuller, Ponderosa High School, CA
Daniel S. Johnson, Silver Creek High School, CA
Brian Lawler, California State University San Marcos, CA
Brent McClain, Vernonia School District, OR
Susan Miller, St. Francis of Assisi Parish School, PA
Amy C. Roszak, Cottage Grove High School, OR
Carmen C. Rubino, Silver Creek High School, CA
Barbara Schallau, East Side Union High School District, CA
Kathleen H. Spivack, Wilbur Cross High School, CT
Wendy Tokumine, Farrington High School, HI

First Edition Multicultural Reviewers
Genevieve Lau, Ph.D., Skyline College, CA
Arthur Ramirez, Ph.D., Sonoma State University, CA
Marilyn Strutchens, Ph.D., Auburn University, AL

Copyeditor
Brandy Vickers

Interior Designer
Marilyn Perry

Production Editor
Andrew Jones

Production Director
Christine Osborne

Editorial Production Supervisor
Kristin Ferraioli

Compositor
Lapiz Digital Services, Kristin Ferraioli

Art Editor/Photo Researcher
Maya Melenchuk

Technical Artists
Lapiz Digital Services, Laurel Technical Services, Maya Melenchuk

Illustrators
Taylor Bruce, Deborah Drummond, Tom Fowler, Briana Miller, Evangelia Philippidis, Sara Swan, Diane Varner, Martha Weston, April Goodman Willy

Cover Designer
Jeff Williams

Printer
Lightning Source, Inc.

Mathematics Product Manager
Elizabeth DeCarli

Executive Editor
Josephine Noah

Publisher
Steven Rasmussen

CONTENTS

Meadows or Malls?

Three-Variable Equations, Three-Dimensional Coordinates, and Matrix Algebra

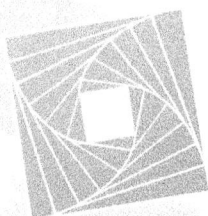

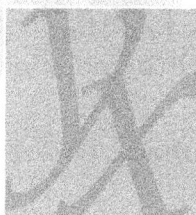

Meadows or Malls?—Three-Variable Equations, Three-Dimensional Coordinates, and Matrix Algebra

Recreation Versus Development: A Complex Problem

The people of River City have a decision to make. They must decide how much of the city's land to use for recreation and how much to use for development, and exactly which land to use for each purpose.

To help them with their decision, you will apply many ideas about algebra, geometry, and the relationship between the two. The first stage is simply to make some sense of this complicated situation.

Daniel Kurek thinks about how to represent the constraints algebraically.

Meadows or Malls?

Who would have thought that so much good fortune could cause so much trouble? Well, it surely has in River City. Actually, there are three separate pieces of good fortune.

First, when Mr. Goodfellow died, he left his 300-acre farm to the city. His will had no stipulations, so the city can do whatever it wishes with the property.

Then the U.S. Army closed its 100-acre base on the edge of town. The federal government gave the land to the people of River City to use in any way they choose.

Finally, 150 acres of city land was leased to a mining company 99 years ago. Now the lease is up. Because the company did not find enough minerals there to make a profit for many years, it does not wish to renew the lease. So that land is also available to River City with no restrictions on its use.

Altogether, that is 550 acres of land the city can use in any way it decides. The problem is that a city isn't exactly an "it." A city is home to many people, who don't always agree. And the people of River City definitely do not agree on how to use the 550 acres.

The controversy centers on two opposing camps. One group wants to use as much of the land as possible for *development*—that is, for stores, businesses, and housing. The other group wants to use as much of the land as possible for *recreation*—that is, for parkland, hiking trails, picnic areas, and a wildlife preserve.

The business community won an initial victory by getting the city council to agree that at least 300 acres will be used for development. The business community proposes that the more attractive sites—the army base and the mining land—should go for development, while any recreation land could come from Mr. Goodfellow's property. But the people with environmental and sporting interests believe that some of the more attractive land should be used for recreation.

continued ◈

The two groups finally arrived at a two-part compromise.

- At most, 200 acres of the army base and mining land will be used for recreation.
- The amount of army base land used for recreation and the amount of farmland used for development together must total exactly 100 acres.

Everyone realizes the city will have to improve any land used for development by putting in sewers, streets, power lines, and so on. The city will also have to spend some money on any land used for recreation.

The city manager made a table listing, for each parcel, how much each type of land use would cost the city. Everyone wants to keep the costs to River City to a minimum.

Parcel	Improvement costs per acre for recreation	Improvement costs per acre for development
Mr. Goodfellow's land	$50	$500
Army base	$200	$2,000
Mining land	$100	$1,000

The city manager has to decide how to split the land use between development and recreation so that the cost of necessary improvements is minimized. She also must ensure that at least 300 acres go for development and that the two-part compromise is followed.

Because of her full schedule, she decides to turn the matter over to a consulting firm of city planners.

continued ▶

Your Task

Your group, functioning as the consulting firm, will work on this problem over the course of the unit. Your task for now is to find out as much as you can about the problem. By the end of the unit, you will identify the best solution.

1. Find one way to allocate the land and satisfy the constraints. Find the cost to the city for this solution (even though you may recognize that it is not the least costly allocation).

2. What approaches to solving this problem might you try if you had more time? What approaches did you try that didn't seem to work?

Meadows, Malls, and Variables

1. Use the variables defined below to write a set of constraints that express the *Meadows or Malls?* problem.

   - G_R is the number of acres of Mr. Goodfellow's land to be used for recreation.
   - A_R is the number of acres of army land to be used for recreation.
   - M_R is the number of acres of mining land to be used for recreation.
   - G_D is the number of acres of Mr. Goodfellow's land to be used for development.
   - A_D is the number of acres of army land to be used for development.
   - M_D is the number of acres of mining land to be used for development.

2. Use the variables from Question 1 to write an algebraic expression for the city's cost based on how the land is allocated.

3. Check whether each of these allocations satisfies the constraints you defined in Question 1. If it does, find the cost to the city. If it does not, show which **constraint** or constraints it violates.

   a. $G_R = 250$
   $A_R = 50$
   $M_R = 150$
   $G_D = 50$
   $A_D = 50$
   $M_D = 0$

   b. $G_R = 200$
   $A_R = 0$
   $M_R = 0$
   $G_D = 100$
   $A_D = 100$
   $M_D = 150$

That's Entertainment!

An entertainer has an ordinary deck of playing cards. He gives them to his subject, turns his back, and has her shuffle the deck thoroughly.

Keeping his back to her so he can't see what she's doing, he then tells her to make some piles according to these instructions.

1. First she turns over the top card of the deck.

   If this is a face card (jack, queen, or king), she puts it back somewhere in the deck and picks the new top card. She keeps going until she gets a card that is not a face card. That is, she continues until she gets an ace, 2, 3, 4, 5, 6, 7, 8, 9, or 10. Then she places that card face up on the table, as the start of a new pile.

2. Beginning with the number on that card, she starts counting to herself until she gets to 12. (Aces are treated as 1.) With each count, she takes one card from the top of the deck and places it face up on top of the new pile. When she reaches 12, she turns the pile over so that the card she started with is face down on top.

   For example, if she initially turns up an 8, she places a card on top of the 8 and silently counts "9." Then she places another card on top of the pile and silently counts "10," then another card on top and counts "11," and finally, another card on top and counts "12." At that point, she turns over the pile, with the 8 face down on top. In this example, the pile would have five cards altogether.

continued ▶

3. As soon as this first pile is complete, she repeats instructions 1 and 2, working with the remaining cards in the original deck. She keeps creating new piles until she runs out of cards.

If she runs out of cards while trying to complete a pile, she picks up all the cards in that incomplete pile.

The subject follows the instructions. When she is done, the entertainer turns around and asks her to give him the cards from her final, incomplete pile.

He sees she has given him five cards, but he does not look to see which cards they are. He also sees she has made six complete piles.

He then tells her to take the top card from each pile and add the values of these cards together, without showing him the cards or telling him the sum.

She does this, and he then tells her the sum she found.

○ Your Task

Your task is to figure out what the sum was and how the entertainer figured it out.

○ Write-up

1. *Problem Statement*
2. *Process*
3. *Solution*
4. *Evaluation*
5. *Self-assessment*

Adapted from *Mathematics: Problem Solving Through Recreational Mathematics*, by Averbach and Chein, Copyright © 1980 by W.H. Freeman and Company. Used with permission.

Meadows or Malls?: Recreation Versus Development: A Complex Problem

9

Heavy Flying

Lindsay is a stunt pilot, but she can't make a living just by doing stunts. So she has bought a transport plane from Philip. He agrees to help her set up her business.

Philip has two customers he no longer has time to serve: Charley's Chicken Feed and Careful Calculators. He suggests that Lindsay deliver their merchandise for them.

Charley's Chicken Feed packages its product in containers that weigh 40 pounds and are 2 cubic feet in volume. Philip has been charging a delivery fee of $2.20 per container.

Careful Calculators packages its materials in boxes that weigh 50 pounds and are 3 cubic feet in volume. Philip has been charging $3.00 per box.

The plane can hold a maximum of 2000 cubic feet of materials and carry a maximum weight of 37,000 pounds.

Charley's Chicken Feed and Careful Calculators can each give Lindsay as much business as she can handle. Of course, she wants to maximize the money she earns per flight so she can spend more time stunt flying.

Your Task

Here is your task for now.

1. Come up with several loads for Lindsay that fit the constraints. Figure out how much she will earn for each load. Assume she charges the same rates Philip did.

2. Use variables and algebra to describe the constraints on what Lindsay can carry.

A Strategy for Linear Programming

River City's problem is somewhat like the Woos' problem in the Year 2 unit *Cookies*. Both are examples of **linear programming** problems, but *Meadows or Malls?* is more complicated. It involves six variables instead of two.

Before tackling this six-variable problem, you will review what you know about two-variable linear programming problems. The goal is to develop a strategy that you might be able to adapt to the more general situation.

Caroline Williams reviews the work she did with two-variable linear programming problems in the Year 2 unit **Cookies.**

Programming and Algebra Reflections

Part I: Programming Reflections

In the Year 2 unit *Cookies,* you learned how to solve two-variable linear programming problems. You've now used ideas and methods from that unit to solve *Heavy Flying.*

To solve *Meadows or Malls?,* which has six variables, you need to generalize those methods. In preparation for creating a generalization that works for more variables, answer these questions to summarize what you know about two-variable problems.

1. What type of information are you given in a two-variable linear programming problem? What are you trying to do?

2. What do you do to solve a linear programming problem in two variables? Describe the process in as general terms as possible.

Part II: Algebra Reflections

Solve each pair of linear equations in two variables algebraically. Explain each step of the process.

3. $4x + y = 13$
   $2x + y = 7$

4. $3x - 2y = 5$
   $x + 3y = 9$

Many types of problems involve finding the common solution to a pair of linear equations. This is the same as finding the coordinates of the point where the graphs of the two equations intersect. You can often estimate the coordinates of this point by sketching the two graphs, but it's helpful to know some algebraic methods that will give you the exact solution.

No single method works best for every **system of equations,** but here are two common approaches. For simplicity, the examples have integer solutions, but the methods work on any system with a unique solution.

The "Setting y's Equal" Method

In this method, you solve both equations to get expressions for one variable in terms of the other. Then you set those expressions equal to each other. Consider this pair of equations.

$$4x - y = 8$$
$$3x + y = 13$$

Adding y and subtracting 8 from both sides of the first equation gives $4x - 8 = y$.

Subtracting $3x$ from both sides of the second equation gives $y = 13 - 3x$.

At the point where the two lines meet, both equations hold true. That is, y is equal to both $4x - 8$ and $13 - 3x$, so $4x - 8 = 13 - 3x$.

Solving this equation gives $x = 3$.

Substituting 3 for x in either $4x - 8$ or $13 - 3x$ gives $y = 4$.

You can check that $x = 3, y = 4$ is a solution of both original equations.

continued ▶

The Substitution Method

In this method, you use one of the equations to express one of the variables in terms of the other and then substitute that expression into the other equation.

Consider this pair of equations.

$$3x + 4y = 18$$
$$2x + y = 7$$

It's easiest to begin with the second equation because the **coefficient** of y is 1. Subtracting $2x$ from both sides gives the equivalent equation $y = 7 - 2x$.

At the point where the graphs of the two original equations intersect, the coordinates must satisfy this new equation. So you can substitute $7 - 2x$ for y into the first equation. In other words, x must satisfy the equation

$$3x + 4(7 - 2x) = 18$$

Solving this equation gives $x = 2$.

Once you know $x = 2$ at the point of intersection, you can substitute 2 for x in the equation $y = 7 - 2x$ to get $y = 3$.

As before, you can check your solution by substituting the values for x and y in the original equations.

Programming Puzzles

1. A Nonroutine Routine

Skateboarder Lance Stunning, who is about to enter a freestyle competition, is busy preparing his routine. His two key moves are the pool drop-in and the rail slide. Lance has to decide how many of each move to include in the routine to maximize his score.

Lance figures he has the stamina to make at most 25 of these moves in his performance. He also knows that his performance can be at most 3 minutes long.

Each pool drop-in takes 4 seconds to perform, and each rail slide takes 9 seconds. If Lance includes more than 20 of either move, the judges will get bored, and he will get a lower score.

Version a: Suppose the judges give each pool drop-in 5 points and each rail slide 10 points. What combination of moves will give Lance the best possible score?

Version b: Suppose the judges give each pool drop-in 7 points and each rail slide 5 points. What combination of moves will give Lance the best possible score?

Version c: Suppose the judges give each pool drop-in 3 points and each rail slide 8 points. What combination of moves will give Lance the best possible score?

continued ▶

2. Planning the Prom

Paige and Remy are organizing the junior prom. They plan to sell two types of tickets: individual tickets and couples tickets. The ballroom where the prom will be held holds a maximum of 400 people.

Paige believes the prom will be more successful if more people go as couples. So that at least half the people at the prom will be in couples, she and Remy decide that the number of individual tickets sold should be at most twice the number of couples tickets.

Remy orders door prizes to be handed out at the prom—one prize for each ticket, even if it's a couples ticket. Unfortunately, the supplier delivers only 225 prizes, and it's too late to get more, so they can sell a total of at most 225 tickets.

Version a: Suppose individual tickets sell for $22 and couples tickets sell for $30. How many of each type of ticket should Paige and Remy sell to maximize the money they take in?

Version b: Suppose individual tickets sell for $15 and couples tickets sell for $35. How many of each type of ticket should Paige and Remy sell to maximize the money they take in?

Version c: Suppose individual tickets sell for $30 and couples tickets sell for $20. How many of each type of ticket should Paige and Remy sell to maximize the money they take in?

continued ▶

3. Working Two Jobs

Raj works as a health aide at both a hospital and a neighborhood clinic. Because he is also going to school part-time, Raj can work no more than 20 hours per week.

Raj likes working at the clinic better than the hospital, so he always wants to work at least as many hours there as at the hospital. However, he wants to keep his job at the hospital in hopes of getting a better position there eventually. To keep the hospital job, he must work there at least 4 hours per week.

Version a: Suppose Raj earns $20 per hour at the hospital and $15 at the clinic. How many hours should he work at each job to maximize his earnings?

Version b: Suppose Raj earns $17 per hour at the hospital and $19 at the clinic. How many hours should he work at each job to maximize his earnings?

Version c: Suppose Raj earns $18 per hour at the hospital and $16 at the clinic. How many hours should he work at each job to maximize his earnings?

Donovan Meets the Beatles

Aji and Sunshine want to make a CD of songs from the 1960s to donate to their school library to be broadcast during school picnics. Sunshine's parents were big fans of the 60s singer Donovan, and they named her after his song "Sunshine Superman." Aji's folks were devotees of the Beatles. Aji and Sunshine want their CD to consist entirely of songs by Donovan and the Beatles.

Because Sunshine is named after a Donovan song, she wants to include at least as many Donovan songs as Beatles songs. Aji agrees. In return, Sunshine agrees to include at least five Beatles songs.

To provide enough variety to make the mix interesting, they decide to include at least 20 songs altogether.

Suppose Aji and Sunshine have to pay $11 in royalties for each Beatles song and $7 for each Donovan song. How many of each type should they put on their CD to minimize their royalty costs?

Finding Corners Without the Graph

When a profit function is linear and the **feasible region** is a polygon, the **profit function** will always achieve its maximum at a corner point of the feasible region. But for problems involving three variables, drawing the feasible region can be difficult. (And it's impossible for more than three variables!)

It's helpful to be able to locate the corner points without actually drawing the region. As preparation for more complex cases, consider the two-variable feasible region defined by these linear inequalities.

$$x + 2y \leq 8$$
$$2x + y \leq 13$$
$$y \leq 3$$
$$x \geq 0$$
$$y \geq 0$$

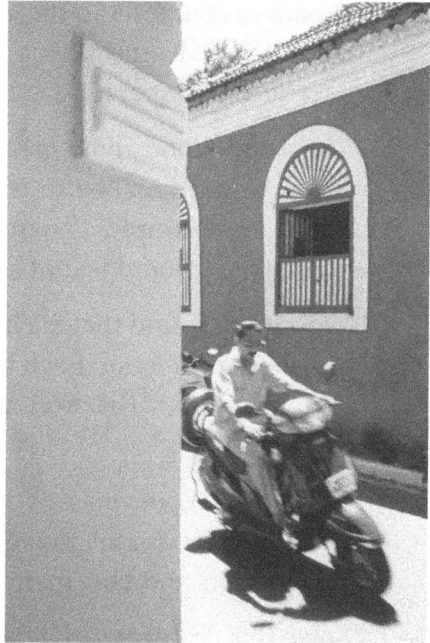

1. Each of these inequalities has a corresponding linear equation, whose graph is a straight line. Each corner point of the feasible region is the intersection of two of these lines. How many combinations of these equations are there, taking them two at a time?

2. For each of your combinations from Question 1, find the intersection point for the pair of lines. If there is no intersection point, explain why not.

3. Which intersection points from Question 2 are actually corner points of the feasible region defined by the inequalities? Explain how you know.

What Wood Would Woody Want?

Woody, a character in the Year 1 unit *Shadows,* is very interested in trees, especially measuring them. Well, he has developed his hobby into a trade and is opening a carpentry shop that makes tables and chairs.

The wood Woody buys is sold in terms of a unit called a *board foot,* which is based on boards with a thickness of 1 inch and a width of 1 foot. For example, a board 9 feet long, 1 inch thick, and 1 foot wide consists of 9 board feet of lumber.

Woody has found that each chair requires 3 board feet of lumber and 2 hours of labor. Each table requires 7 board feet of lumber and 8 hours of labor. His profit on each chair is $15, and his profit on each table is $45.

This week, Woody has 420 board feet of lumber and 400 hours of labor available. (He doesn't do all the work himself.) He wants to know how many chairs and how many tables he should make to maximize his profit.

Solve this problem using the general strategy for working on linear programming problems without drawing a feasible region. Use C for the number of chairs Woody makes and T for the number of tables. Explain your work carefully. If you discover places where the strategy is unclear or doesn't seem to work correctly, make a note of them.

Adapted with permission from the *Mathematics Teacher,* © May 1991, by the National Council of Teachers of Mathematics.

Widening Woody's Woodwork

Consider a variation on the situation from *What Wood Would Woody Want?* Assume as before that each chair requires 3 board feet of lumber and 2 hours of labor and that each table requires 7 board feet of lumber and 8 hours of labor.

Suppose Woody has expanded his operations so that he has 630 board feet of lumber and 560 hours of labor available each week. Also suppose he has changed his prices so that his profit on each chair is now $18 and his profit on each table is now $42.

How many chairs and how many tables should Woody make to maximize his profit?

More Equations

Part I: Pairs of Equations

In working on linear programming problems, you often need to solve pairs of linear equations. Use the **substitution method** or another algebraic method to try to solve each pair of equations. Show your work.

1. $5x + 3y = 7$ and $y = x - 3$

2. $3x + 2y = 11$ and $x + y = 4$

3. $5x - 3y = 5$ and $10x + 6y = 20$

4. $2x - 3y = 2$ and $4x - 6y = 9$

5. $2x + 4y = 12$ and $6x + 12y = 36$

Part II: Look It Up

The set of points in the xy-coordinate system is often referred to as the **coordinate plane.** You will see that planes play an important role in this unit.

The word *plane* has a specific meaning in geometry, but it has other meanings in different contexts. Look up this word in the dictionary. Write down as many meanings for it as possible, including the geometric definition.

Equations, Points, Lines, and Planes

You have solved two-variable linear programming problems by looking at the intersections of lines using the xy-coordinate system. You will eventually return to River City and its land-use problem. But first you need to move to another level of complexity for graphs.

In the next portion of this unit, you will develop a coordinate system for three variables and see what the graph of a linear equation is in this new setting.

You will also look at how lines and planes intersect in order to generalize what you know about how lines intersect in the plane.

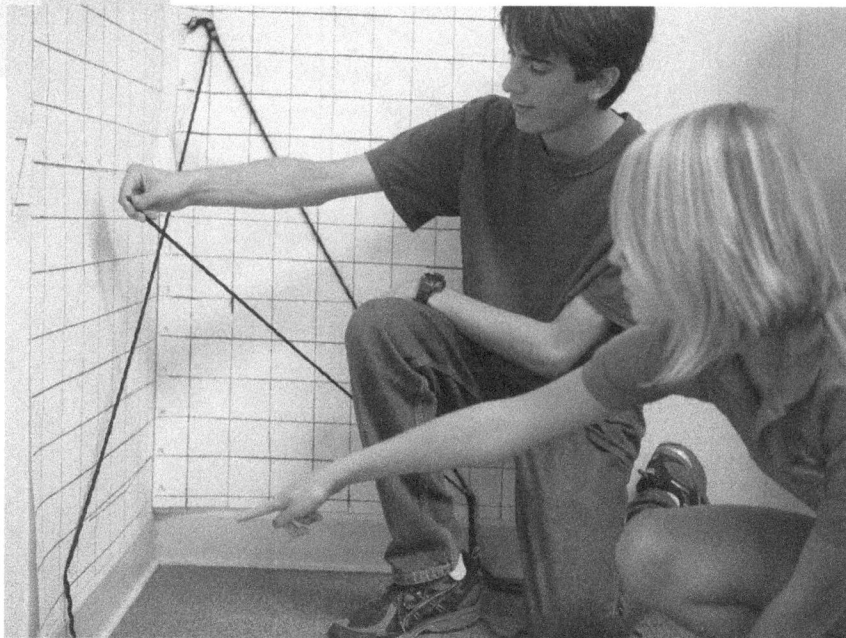

Gavilan Galloway and Catherine Borror create a three-dimensional coordinate system.

Being Determined

1. Do two lines uniquely determine a point? In other words, if l_1 and l_2 are lines in a plane, is there always one, and only one, point that lies on both of them?

   Explain your answer. In particular, if there are exceptions, state what they are and what happens in the exceptional cases. (Remember, in mathematics the word *line* always refers to a straight line, which does not have to be vertical or horizontal.)

2. Do two points uniquely determine a line? In other words, if P and Q are points in a plane, is there always one, and only one, line that goes through both P and Q?

   Explain your answer. In particular, if there are exceptions, state what they are and what happens in the exceptional cases.

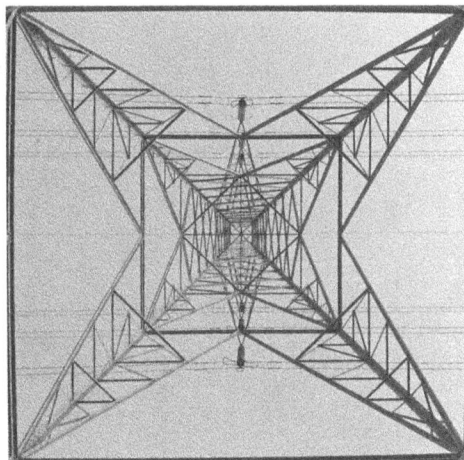

How Much After How Long?

1. The performing arts department put on its spring show on Friday and Saturday nights. The price of a ticket was the same both nights, and the cost of putting on the show was also the same both nights.

   The department made a profit of $400 on Friday when 100 people bought tickets and a profit of $500 on Saturday when 120 people bought tickets.

   a. How much did one ticket cost? What was the cost of putting on the show each time?

   b. Let *t* represent the number of people who buy tickets, and let *p* represent the amount of profit the department would make from selling that many tickets. Think of *p* as a function of *t*. Find a rule for this function, expressing *p* in terms of *t*.

2. Joey has saved some money, but he decides to get a job and add his earnings to his savings so he can buy a car. After working 10 hours, he has a total of $210. After working 120 hours, he has $870 altogether.

   a. How much money did Joey make per hour? How much money did he already have before he started working?

   b. Let *h* represent the number of hours Joey has worked, and let *m* represent the amount of money he has accumulated altogether after *h* hours. Think of *m* as a function of *h*. Find a rule for this function, expressing *m* in terms of *h*.

continued ◗

3. In Question 1, you had two combinations of ticket sales and profit.
   - Selling 100 tickets produced a $400 profit.
   - Selling 120 tickets produced a $500 profit.

   a. What were the two similar combinations in Question 2?

   b. In each question, two combinations is enough information to allow you to find a rule that describes the situation. What does this have to do with the activity *Being Determined?*

4. A line passes through the points (2, 15) and (7, 45). Find an equation for the line, and explain how you got your answer.

The Points and the Equations

As you know, if P and Q are two distinct points in the plane, then there is a unique line that goes through them both. If the points are given as coordinate pairs in the xy-plane, you may want to look for an equation whose graph is that line through the points.

1. For each pair of points, find a linear equation whose graph will go through the two points. Use x to represent the first coordinate and y to represent the second coordinate.

    a. $(4, 9)$ and $(6, 13)$

    b. $(5, 13)$ and $(3, 7)$

    c. $(8, 8)$ and $(20, 14)$

    d. $(-2, 5)$ and $(1, -4)$

2. For each pair of points in Questions 1a and 1b, create a word problem that would require the solver to find the equation. Use Questions 1 and 2 of *How Much After How Long?* as sources of ideas for situations, or create your own situations.

The Three-Variable Coordinate System

The familiar *xy*-coordinate system is used to represent pairs of numbers by points in the plane. For example, the diagram to the right represents the combination of values $x = 2$ and $y = -3$.

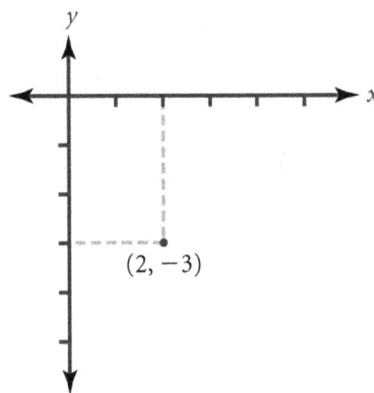

A similar system can be used to represent triples of numbers. One common way to do this is to picture the *x*-axis and *y*-axis as "lying flat" and the *z*-axis as "coming out" perpendicular to that plane. It's difficult to represent this system in two dimensions, but the diagram below suggests one way to do this.

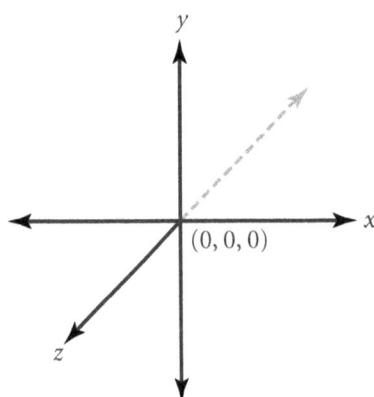

We consider the positive direction of the *z*-axis to be coming out of the page. The dashed line represents the negative portion of the *z*-axis.

As in the two-variable system, the point where the three axes meet is called the *origin*. It represents the values $x = 0$, $y = 0$, and $z = 0$. We write this point as $(0, 0, 0)$.

A triple of values represents moving the appropriate distances in the appropriate directions from the origin.

continued

For instance, the point $(2, -3, 4)$ is found, as shown in the diagram at the right, by going 2 units to the right of the origin, 3 units down, and 4 units "toward you." This point represents the values $x = 2$, $y = -3$, and $z = 4$.

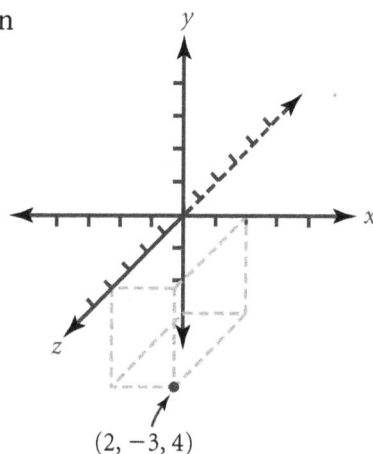

The collection of all points in this system is called the **three-dimensional coordinate system.** It is often referred to simply as **3-space.** Because our world is three dimensional, this system is very useful for describing real-world phenomena, such as the position of an object in space.

$(2, -3, 4)$

Each pair of axes defines a plane, and these planes are known as the *coordinate planes.*

In the diagram at the right, the light shaded plane is called the *xy-plane,* the white plane is called the *xz-plane,* and the darker shaded plane is called the *yz-plane.*

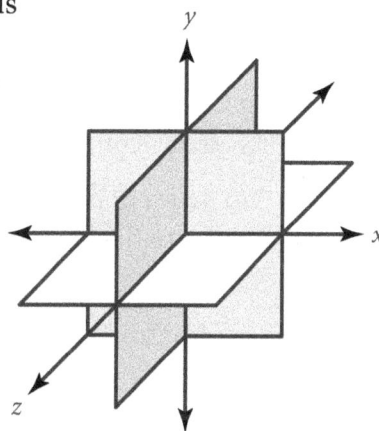

These planes divide 3-space into eight separate regions, known as **octants.** The octants are analogous to the quadrants of the two-variable coordinate system.

Although there is no standard numbering system for all the octants, the set of points whose coordinates are all positive is called the **first octant.**

What Do They Have in Common?

In the two-variable coordinate system, the points in the first quadrant all have positive *x*-coordinates and positive *y*-coordinates.

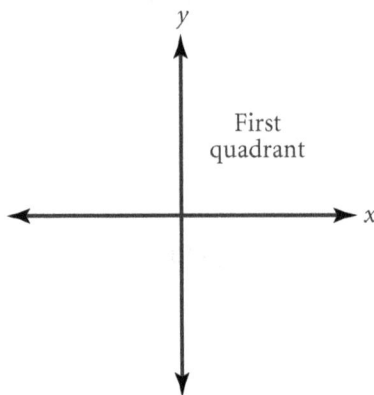

Similarly, there are sets in the three-dimensional coordinate system whose points all have one or more characteristics in common.

Questions 1 to 8 describe various sets in geometric terms. The descriptions refer to the diagram shown below. For each set, do two things.

- Give the coordinates of five specific points in the set.
- State what characteristic or characteristics the points in the set have in common, in terms of their coordinates.

1. The set of points in the *yz*-plane
2. The set of points in the *xz*-plane
3. The set of points on the *x*-axis
4. The set of points on the *z*-axis
5. The set of points above the *xz*-plane
6. The set of points behind the *xy*-plane
7. The set of points 1 unit to the left of the *yz*-plane
8. The set of points in the first octant

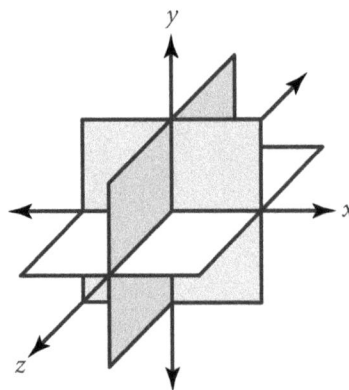

Trying Out Triples

You have been looking at the graphs of some special linear equations in three variables—namely, examples in which not all three variables appear.

Soon you will be graphing the equation $x + y + \frac{1}{2}z = 4$. This linear equation involves all three variables. As preparation, answer these questions, which concern the equation $x + y + \frac{1}{2}z = 4$.

1. Find a dozen or so triples that are solutions to this equation.

2. Organize your solutions in some way that you think might be helpful in finding the graph of this equation.

3. Describe what you think this graph will look like, or make a sketch or model.

More Cookies

Abby and Bing Woo were the bakery owners in the main problem from the Year 2 unit *Cookies*.

The Woos still make plain and iced cookies. But now, by popular demand, they're also selling chocolate chip cookies.

Here's a reminder of the ingredients needed for the original cookies.

- One dozen plain cookies requires 1 pound of cookie dough (and no icing).
- One dozen iced cookies requires 0.7 pound of cookie dough and 0.4 pound of icing.

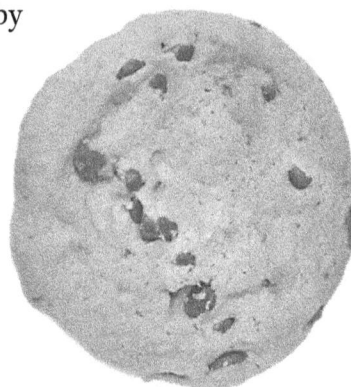

Now you need the same information for the chocolate chip cookies.

- One dozen chocolate chip cookies requires 0.9 pound of cookie dough and 0.15 pound of chocolate chips (and no icing).

The Woos are limited by the ingredients they have on hand.

- 120 pounds of cookie dough
- 32 pounds of icing
- 18 pounds of chocolate chips

In the past, the Woos were also limited by the amount of work time available. They now have other family members to help, so work time is no longer a limitation on how many cookies they can make. They have also bought more ovens, so oven space is no longer a limitation either.

Plain cookies sell for $6.00 a dozen and cost $4.50 a dozen to make. Iced cookies sell for $7.00 a dozen and cost $5.00 a dozen to make. Chocolate chip cookies sell for $10.00 a dozen and cost $7.75 a dozen to make.

The Woos want to know how many dozens of each kind of cookie to make to maximize their profit.

continued ◗

Eventually you will need to answer the Woos' question. For now, do this.

1. Write a set of constraints that express the situation just described using these variables.

   - P represents the number of dozens of plain cookies.
   - I represents the number of dozens of iced cookies.
   - C represents the number of dozens of chocolate chip cookies.

2. Find four different combinations of cookies the Woos can make. Compute the total profit for each combination.

Just the Plane Facts

You have seen that two lines in a plane usually uniquely determine a point. That is, except for the special cases of parallel lines or two lines that are identical, two given lines will intersect in one and only one point.

Your task in this activity is to examine other cases of "determining," this time in 3-space. In each case, do two things.

- Describe the different ways in which the given objects can intersect.
- Explain your answers with diagrams, three-dimensional models, or other appropriate devices.

1. A line and a plane

2. Two planes

3. Three planes

4. Two lines (*Note:* Your work in Question 1 of *Being Determined* concerned lines in the same plane. Now think about the more general situation, in which the lines may or may not be in the same plane.)

5. Four planes

6. Any other combinations you'd like to investigate

Solving with Systems

You may be able to find the answers to this activity without using algebra. However, to help develop your algebraic skills for work with more difficult problems, you should define variables, write a system of equations, and use substitution to solve the systems for each question.

1. Ming is a competitive surfer. The two moves she used in her most recent competition are the off-the-lip and the cutback. The wave she caught allowed her to do a total of six moves.

   The judges awarded 6 points for each off-the-lip move and 8 points for each cutback. Ming scored a total of 40 points.

   How many moves did she make of each type?

2. In the activity *Programming Puzzles*, Paige and Remy sold two types of prom tickets: individual tickets and couples tickets.

   Suppose they sold individual tickets for $10 and couples tickets for $18 and collected $1,500 altogether. If 160 people attended the event, how many tickets of each type did they sell? (*Caution:* Remember that a couples ticket represents two people.)

Fitting a Line

Although a given line contains many points, a given *pair* of distinct lines in a plane determines a *unique* point, except when the lines are parallel. Similarly, although many lines go through a given point, a pair of distinct points always determines a unique line.

In the activity *Being Determined,* you looked at these ideas from a geometric point of view. Now you will use algebra to find the equation for a line that goes through two specific points in a plane.

Some Lines Through (1, 2)

First consider the point $(1, 2)$. You can see that this point lies on the graph of the linear equation $y = 5x - 3$ by substituting 1 for x and 2 for y. Informally, we say that the line $y = 5x - 3$ "goes through" the point $(1, 2)$.

1. Show that the line $y = -5x + 7$ also goes through $(1, 2)$.

2. Find equations for two other lines that go through $(1, 2)$.

The Family of Lines Through (1, 2)

As the diagram suggests, infinitely many functions of the form $y = ax + b$ have graphs that go through $(1, 2)$. But exactly which linear functions are they? What values of a and b give lines through this point?

For instance, for the function $y = 5x - 3$, we have $a = 5$ and $b = -3$, so these values for a and b give a line through $(1, 2)$.

3. a. Find the values of a and b for the equation $y = -5x + 7$.

   b. Find the values of a and b for each equation you found in Question 2.

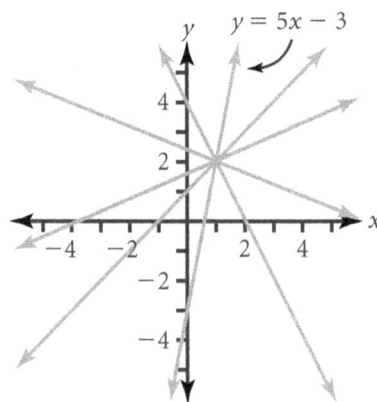

continued ◗

c. Look for a relationship between a and b that holds for all the lines that go through the point $(1, 2)$. Express this relationship as a linear equation involving a and b.

Suggestion: If you are having trouble finding a relationship, find the equations for some more lines that go through $(1, 2)$ and compile a table of values for a and b. Then look for a pattern in your table.

The Family of Lines Through $(-1, -6)$

Now consider a second point, $(-1, -6)$. Again, as the diagram suggests, infinitely many linear functions have graphs that go through this point.

4. a. Find three examples of equations of the form $y = ax + b$ that go through this point.

   b. Find a linear equation involving a and b that must hold for any line $y = ax + b$ that goes through $(-1, -6)$. You might use a method like that suggested in Question 3c.

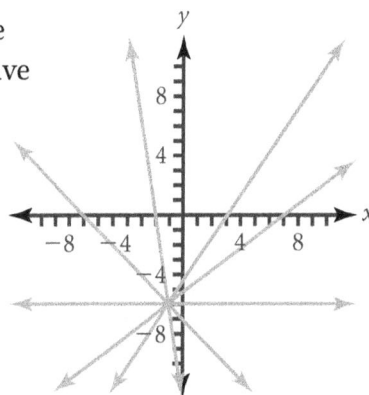

A Line Through Both Points

Now suppose you want a line $y = ax + b$ that goes through both points.

5. Find values for a and b so that the line $y = ax + b$ goes through both $(1, 2)$ and $(-1, -6)$.

Cookies, Cookies, Cookies

More Cookies introduced you to a linear programming problem in three variables involving the Woos' bakery. In that activity, you simply wrote the constraints and found some possible cookie combinations the Woos could make.

Now you're ready to solve that problem. To do so, you will use what you've learned about planes, linear equations in three variables, and the general strategy for solving linear programming problems.

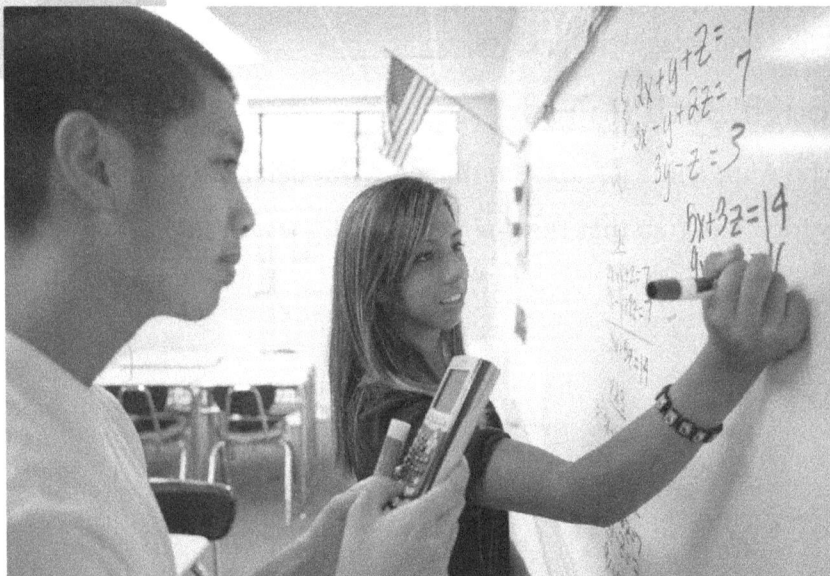

Phi Nguyen and Kelsey Coria solve a linear programming problem.

SubDivvy is a number game for two players. Here are the rules.

1. Two players cooperatively choose a starting number greater than 1. We'll call that number N.

2. Player 1 chooses a positive divisor of N that is different from N—that is, a number that divides "evenly" into N, so that the remainder is zero. Player 1 subtracts that divisor from N and gives the result of that subtraction to Player 2.

3. Player 2 works with the result of the subtraction just as Player 1 worked with N. That is, Player 2 chooses a positive divisor of that number, different from the number itself, and subtracts the chosen divisor from it. Player 2 gives the result of this subtraction to Player 1.

4. Players take turns choosing divisors and subtracting until the result reaches the number 1. The player who produces the result of 1 is the winner.

Here is a sample game, with explanations.

68	The number 68 is selected as the starting number N.
−4	Player 1 chooses 4, which is a divisor of 68, and subtracts.
64	The result of the subtraction is 64, which is given to Player 2.
−16	Player 2 chooses 16, which is a divisor of 64, and subtracts.
48	The result of the subtraction is 48, which is given to Player 1.
−24	Player 1 chooses 24, which is a divisor of 48, and subtracts.
24	The result of the subtraction is 24.
−8	Player 2 chooses 8, which is a divisor of 24, and subtracts.
16	The result of the subtraction is 16.
−8	Player 1 chooses 8, which is a divisor of 16, and subtracts.
8	The result of the subtraction is 8.
−4	Player 2 chooses 4, which is a divisor of 8, and subtracts.
4	The result of the subtraction is 4.

continued

-2	Player 1 chooses 2, which is a divisor of 4, and subtracts.
2	The result of the subtraction is 2.
-1	Player 2 chooses 1, which is a divisor of 2, and subtracts.
1	The result of the subtraction is 1, so the game ends. Player 2 is the winner.

Your POW is to explore this game.

One important task is to investigate the issue of who wins. Examine whether there are certain starting numbers for which Player 1 has a winning strategy—that is, a complete strategy by which Player 1 can win no matter what Player 2 does on any turn. Are there starting numbers for which Player 2 has a winning strategy? Are there starting numbers for which no player has a winning strategy?

Consider other questions besides the issue of who wins. For instance, for different starting numbers, what can you say about the shortest game possible? The longest game possible? What generalizations can you make?

What are some other questions you might investigate about this game?

o *Write-up*

1. *Process*

2. *Solution*

3. *Evaluation*

4. *Self-assessment*

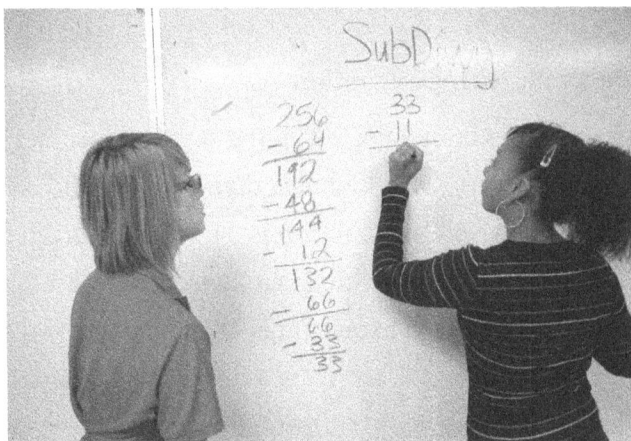

Adapted from Fendel/Resek, *Foundations of Higher Mathematics: Exploration and Proof*, content from pp. 6–7, © 1990 Addison-Wesley Publishing Company Inc. Reproduced by permission of Pearson Education, Inc.

The *More Cookies* Region and Strategy

The activity *More Cookies* introduced you to a more complex version of the original problem from the *Cookies* unit. We will refer to the situation in that activity as "the *More Cookies* problem."

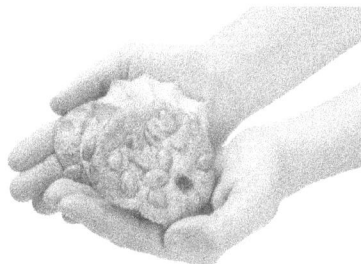

1. You have already developed a set of constraints for the *More Cookies* problem and seen that they are linear inequalities in three variables. You have also learned about setting up a coordinate system for equations involving three variables. In particular, you now know that the graph of a linear equation in three variables is a plane.

   Using that knowledge, give a general description of what the feasible region for the *More Cookies* problem should look like. For a challenge, you might sketch or build a model of this region.

2. Earlier in this unit, you developed a general strategy for solving two-variable linear programming problems. One element of that strategy was to identify corner points of the feasible region by finding the intersections of pairs of linear equations corresponding to the constraints.

   Explain how you could adapt that part of the strategy to work for linear programming problems involving three variables.

Finishing Off the Cookies

Your group will now develop a general plan for solving the *More Cookies* problem—that is, for finding the number of dozens of each type of cookie that will maximize the Woos' profit.

In your group, decide on an assignment for each group member so that the group can quickly solve the *More Cookies* problem together. You may decide that each person should do a different piece of the problem, or you may have several people doing the same thing as a check. It's up to you.

Your written work for this activity has three parts.

1. State your group's general plan.

2. State your individual assignment as part of that plan.

3. Describe what you did and what conclusions you reached for your part of the plan.

Equations, Equations, Equations

You've seen that solving a linear programming problem involves solving systems of linear equations.

The *More Cookies* problem has only three variables, and only one of the equations uses more than one variable. To solve the River City land-use problem, you will need to learn much more about solving such systems of equations. That's your task in the next portion of the unit.

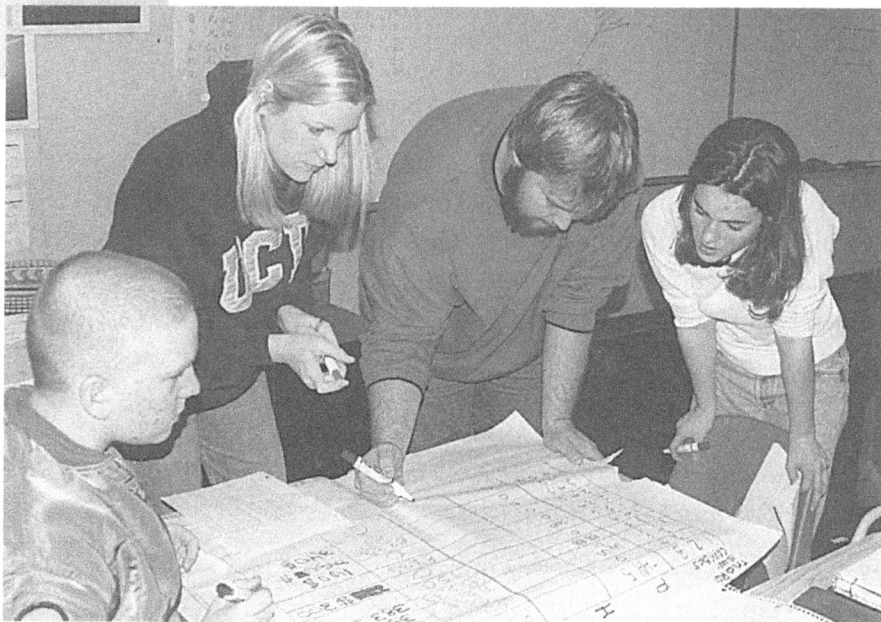

Completing a three-variable linear programming problem will aid Randy Stevens, Emily Gubser, Kyle Abraham, and April Long in solving the more complex River City land-use problem.

Easy Does It!

To solve the *More Cookies* problem, you needed to find the common solution to some systems of three linear equations in three variables.

Those systems of equations were mostly simple ones, but real-life problems aren't always so straightforward. So here's a chance to learn more about solving such systems. (For these questions, assume there is no sales tax involved.)

1. Consider this question.

   Will bought 4 packages of batteries and 1 package of CDs for $20.00. Tania bought 5 packages of batteries and 1 package of CDs for $23.00. How much does a package of batteries cost?

   You can probably answer this question without using equations, given that Will and Tania made almost the same purchases. But write down a pair of equations anyway, and explain how your intuitive reasoning about the question could be expressed in terms of the equations.

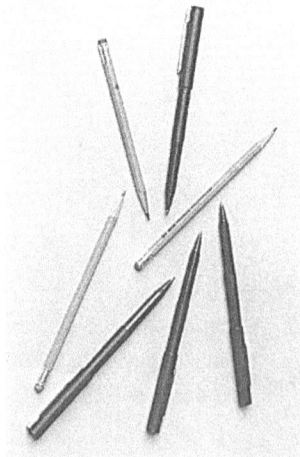

2. Here is a similar question.

   Jennie bought 4 pens and 3 pencils for $3.75. Tanisha bought 4 pens and 6 pencils for $4.50. How much does a pencil cost? How much does a pen cost?

   Again, you can probably answer this question without using equations, but write down a pair of equations anyway. Explain how your intuitive reasoning about the question could be expressed in terms of the equations.

continued

3. This question is not quite as easy as Question 2, but the same sort of approach works fairly well.

Sanji bought 5 pears and 3 apples for $5.80. Ursula bought 10 pears and 7 apples for $12.20. How much does an apple cost?

Write down a pair of equations that express the situation. Explain how your intuitive reasoning about the question could be expressed in terms of the equations.

Get Rid of Those Variables!

Read the situation described here and then move on to the questions. The situation probably sounds similar to the questions in the activity *Easy Does It!*, but be sure to follow the instructions carefully.

> Erin bought 3 bottles of juice and 7 pounds of carrots and spent $12.90. Jinho bought 5 bottles of juice and 6 pounds of carrots and spent $14.70.

1. Represent the information in the situation as a pair of linear equations.

2. a. Generate a list of at least ten more combinations of purchases whose cost is easy to figure out from the combinations given in this situation. (Don't figure out the price of a bottle of juice or a pound of carrots yet.)

   b. Represent each combination from your list, with its cost, using an equation.

3. Find a pair of equations in your list that makes it easy to see what the price of a bottle of juice or a pound of carrots is. Explain how to find these prices from your pair of equations.

   If you don't find such a pair, create more combinations, until getting the individual prices is as simple as it was in *Easy Does It!*

Eliminating More Variables

Whenever you have a system of linear equations, you can create new equations by adding or subtracting two of those equations or by multiplying an equation by a **constant.**

If the coefficients match up appropriately, then adding or subtracting equations makes one of the variables "disappear," giving an equation with one less variable.

This technique for solving systems of linear equations is called the **elimination method** (or *Gaussian elimination*).

Solve each pair of equations using the elimination method. Be sure to show your work clearly.

1. $2a + 3b = 8$
   $4a + 9b = 22$

2. $4r + 3s = 18$
   $r + 2s = 7$

3. $2w + 3z = 10$
   $5w + 7z = 17$

4. $3p + 4q = 10$
   $5p - 4q = 6$

5. $-2c + 5d = 9$
   $3c + 2d = 15$

6. $2f - 6g = -16$
   $3f + 5g = 25$

Gardener's Dilemma

Part I: Leslie Returns

Leslie the landscape architect, whom you encountered in *Orchard Hideout,* is back to ask for your help with another problem.

There has been a drought this summer. Leslie would like to tell her clients how much water they will need for their gardens.

She has divided the plants she uses into three general categories: lawns, flowers, and shrubs. Her goal is to determine how much water per square foot each category of plant requires.

She looked up the water-usage records for three families she worked for during the last drought. She knows that the amounts of water these families used at that time for each type of plant were adequate without being wasteful.

The table below shows the weekly water usage of the three families. Assume the amount of water used per square foot of lawn, flowers, and shrubs is the same for each family.

	Number of square feet of lawn	Number of square feet of flowers	Number of square feet of shrubs	Number of gallons of water used
Family 1	900	120	40	1865
Family 2	0	160	800	180
Family 3	120	80	240	310

Define variables and write a system of equations that could be used to find the amount of water needed for a square foot of each category of plant. You do not need to solve this system of equations.

continued

Part II: Elimination

Solve each system of equations using the elimination method.

1. $6c + 5d = 2$
   $2c + 7d = 22$

2. $5u + 9v = 13$
   $4u - 5v = 47$

3. $4x - 9y = -95$
   $18x + 20y = 480$

4. $-5r + 2s = -7$
   $3r - 4s = 3$

Elimination in Three Variables

Leslie's task in *Gardener's Dilemma* can be expressed using a system of three linear equations in three variables.

The elimination method you have used for two-variable systems can be applied to three-variable systems. The extra variable, though, makes things a bit more complicated.

Because the coefficients in Leslie's equations are fairly large, that system isn't the best one for learning the method. Here are some simpler systems involving three linear equations and three variables. Solve each of them using the elimination method.

1. $4x + 2y + z = 9$
   $2x - y - z = 2$
   $x + 2y + 3z = 9$

2. $3u + v + w = 9$
   $u + v - w = 5$
   $u + 2v + w = 4$

3. $5r - s + 3t = -10$
   $-2r + 2s + t = 11$
   $r + s + t = 2$

More Equation Elimination

In this activity, you will continue your work with the elimination method. Try to use that approach to solve each of these systems of linear equations in three variables.

1. $3a + 2b + 3c = 2$
   $2a + 5b - c = -3$
   $3b = -6$

2. $3u - v + 3w = 3$
   $2u - 4v + 5w = -10$
   $u + v + w = 1$

3. $x + y + z = 3$
   $x - y = 5$
   $y - z = -7$

4. $2d - e + f = 5$
   $d + 2e + f = 3$
   $3d + e + 2f = 8$

Equations and More Variables in Linear Programming

What do you do if one of your linear programming constraints is an equation instead of an inequality? What do you do if you have too many variables to make a graph?

In the next several activities, you will focus on two new linear programming problems: *Ming's New Maneuver* and *Eastside Westside Story*. The first problem addresses the issue of constraint equations. The second extends all of the ideas you have learned about linear programming to four variables.

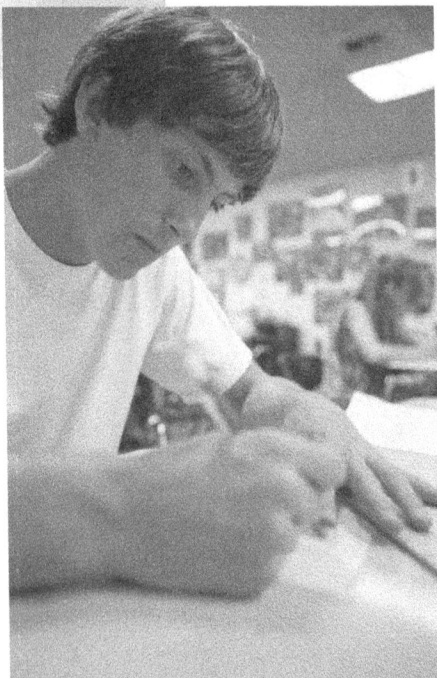

Kevin Brandt refines his strategy for solving linear programming problems.

Ming's New Maneuver

Ming, from the activity *Solving with Systems*, has finally perfected the tube ride! She wants to use it along with her off-the-lip moves and cutbacks in her next competition.

The waves at this new beach are well suited for the tube ride, but Ming cannot do the same move over and over. She still needs to plan her ride to maximize the points she earns.

In this competition, Ming must make exactly 20 moves. Also, because a lip-gloss company is sponsoring the event, she is required to do at least 3 off-the-lip moves.

Ming decides she should use at most 24 seconds for her tube rides and cutbacks. She estimates the cutbacks will take 1 second each and the tube rides will take 2 seconds each.

Here is the scoring system the judges have announced.

- Off-the-lip moves: 1 point each
- Cutbacks: 4 points each
- Tube rides: 5 points each

1. Set up a system of constraints for this situation. Use L for the number of off-the-lip moves, C for the number of cutbacks, and T for the number of tube rides.

2. Determine how many of each type of move Ming should perform to maximize her point total.

Let Me Count the Ways

In solving the *More Cookies* problem, you needed to consider all combinations of constraints taken three at a time. (You may have realized that some of the resulting systems had no solution, so you could eliminate them immediately.)

This activity illustrates other situations in which you might want to make complete lists.

1. Paula's favorite pizza place is offering a special price for ordering exactly two toppings on a pizza. The two toppings have to be different.

   The store has eight toppings to choose from.

   - Anchovies
   - Onions
   - Mushrooms
   - Sausage
   - Olives
   - Zucchini
   - Peppers
   - Pineapple

   Make a complete list of all the combinations Paula can choose from for her two-topping pizza. (You may recall a similar problem from the Year 1 unit *The Game of Pig*.) It doesn't matter in what order you list the two choices. For instance, "anchovies and mushrooms" is the same choice as "mushrooms and anchovies."

continued

2. Fraser School goes from seventh through twelfth grade. A Student Advisory Group has been elected that consists of six students, one from each grade level.

   a. The Student Advisory Group needs to pick a three-person committee from among their members to help plan the homecoming dance. Make a complete list of all the possible combinations of members for this committee.

   b. Suppose the group decides that the student representing the twelfth grade should definitely be on the committee, because this will be that class's last homecoming dance. List all the possible committees now.

3. Jared works afternoons at the local grocery store. He's been an employee long enough that he gets to choose which four afternoons he will work each week. The store is open seven days a week.

   Make a complete list of all his possible choices.

Three Variables, Continued

You know that the graph of a linear equation in three variables is a plane in 3-space. You also know that the intersection of three planes can be

- nothing
- a single point
- infinitely many points (either a line or a plane)

Each of these questions gives a system of three linear equations, so the graphs of the three equations are three planes. For each system, state which of the three types of intersection the graphs have, and justify your answer. If the intersection is a single point, find that point.

1. $a - b + 2c = 2$
   $2a + 2b - c = -3$
   $3a + b + c = 4$

2. $r + s + t = 2$
   $2r + 2s + 2t = 4$
   $3r + 3s + 3t = 6$

3. $u - v + w = 2$
   $u + v + w = 6$
   $u + v + 2w = 9$

4. $2x = 6$
   $3y + z = -7$
   $6x + 6y + 2z = 4$

Grind It Out

Solving systems of linear equations is an important part of various kinds of problems, including linear programming problems.

As the number of variables grows, the systems often become more difficult to solve. The central unit problem, *Meadows or Malls?,* involves six variables. This activity gets you a bit closer to that level of complexity.

1. Solve this system of four linear equations in four variables.

$$2x + y - 3z + w = 6$$
$$x + y + 2z + w = -1$$
$$y - z + w = 0$$
$$x + z - w = 5$$

2. You know that two distinct lines usually intersect in a single point, so a system of two linear equations in two variables usually has a unique solution. Similarly, as you saw in *Just the Plane Facts,* three planes in 3-space will usually have a unique point of intersection, so a system of three linear equations in three variables usually has a unique solution.

   For more than three variables, it isn't possible to draw diagrams or build models to see what's happening geometrically. Nevertheless, a system of n linear equations in n variables usually has a unique solution.

   Give the best explanation you can for why a system of four linear equations in four variables should usually have a unique solution.

Crack the Code

People have devised many kinds of secret codes to be able to communicate privately without others understanding their messages.

○ Letter-Substitution Codes

Many messages involve words. One of the most popular ways to encode a word message is to substitute a different letter for each letter of the alphabet.

If the person who receives your message knows your system for replacing letters, it is easy to figure out your code. Even if that person does not know your system, it may not be too difficult to figure out your message, because of certain special letter combinations and the frequency with which certain letters occur.

○ A Letter-Number Code

This POW concerns codes for arithmetic problems rather than word messages. To use such a code, you start with an arithmetic problem such as

$$\begin{array}{r} 35 \\ + \, 35 \\ \hline 70 \end{array}$$

To create a coded version of the problem, you replace each number with a letter, always using the same letter for a particular number. For example, you might replace 3 with A, 5 with D, 7 with O, and 0 with H. The addition problem then becomes

$$\begin{array}{r} A\,D \\ + \, A\,D \\ \hline O\,H \end{array}$$

In using such codes, be careful to distinguish between the number 0 and the letter O.

continued ▶

○ *Figuring Out the Code*

It's easy to make up such a code, and it's also easy to figure out what the coded problem represents if you know the replacement system.

What's more interesting is trying to figure out a code merely by looking at the coded problem. That is, you are shown only the problem written with letters, and you have to figure out the original arithmetic problem.

○ *The Rules*

Problems like these usually follow certain rules.

• If a letter is used more than once in the same problem, it stands for the same number each time.

• Different letters in the same problem always stand for different single-digit numbers.

• A letter standing for 0 never starts a number with more than one digit. For example, the final arithmetic problem can't have a number like 05, but it can use 507 or 80 or even simply 0.

For some letter problems, it is easy to reconstruct the original arithmetic problem. For others, it can be quite difficult. Sometimes there is no possible answer, and sometimes there are many possible answers.

○ *The Problems*

See whether you can crack the codes for these problems based on the rules just listed. If you think there is only one right answer, prove it. If you think there are several possibilities, give them all and prove there are no others. You will need to keep careful track of how you arrive at your answers.

$$1. \quad \begin{array}{r} A\,B\,B \\ -\quad\ A \\ \hline D\,D \end{array}$$

$$2. \quad \begin{array}{r} S\,S \\ +\,E\,E \\ \hline S\,S\,T \end{array}$$

$$3. \quad \begin{array}{r} A\,B \\ +\,B\,C \\ \hline A\,D\,E \end{array}$$

continued ▶

4. This one is definitely harder than the others.

$$\begin{array}{r} S\ E\ N\ D \\ +\ M\ O\ R\ E \\ \hline M\ O\ N\ E\ Y \end{array}$$

5. Make up a problem of your own that has a unique solution, and prove the solution is unique.

○ *Write-up*

1. *Process* and *Solution:* Do a separate write-up for each of Questions 1 to 5, combining the process and solution components for each problem. You must *prove* your solutions are the only ones possible. You may find that explaining the process you went through to decipher the code will be part (or perhaps all) of your proof.

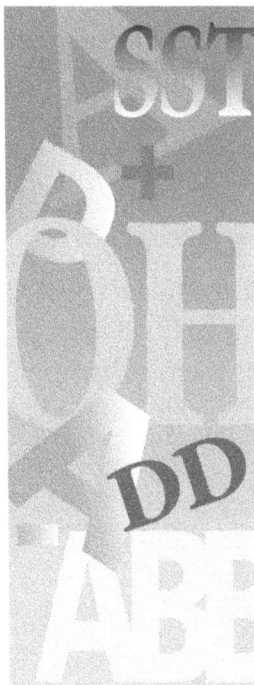

2. *Self-assessment*

Constraints Without a Context

This system of linear constraints in three variables defines a feasible region.

    a. $x + 2y + z \leq 120$

    b. $3x + 2z \leq 12$

    c. $y + z = 8$

    d. $x \leq 10$

    e. $y \leq 7$

    f. $x \geq 0$

    g. $y \geq 0$

    h. $z \geq 0$

1. Create a list of systems of equations you could solve so that the solutions to those systems would include all the corner points of this feasible region. If you can see that a particular system does not have a solution, you may omit it from your list, but you must explain how you know it has no solution.

2. Explain how you would use your list from Question 1 to determine the actual corner points of the feasible region.

Eastside Westside Story

For many years, River City had only one high school. Known simply as River High, it is located on the west side of the river. But as the city grew, the school became quite crowded.

Finally the community built a new school, on the east side of the river. They call it New High.

To promote the idea that both schools serve the entire city, the school board has mandated that at least half of the students attending New High should come from the west side of the river.

The school district has always used buses to bring students to River High. Now the district needs to provide two sets of buses. It is anxious to minimize busing costs.

Here are some facts about the situation.

- There are 300 high school students living on the east side of the river and 250 living on the west side.
- New High can handle up to 350 students, and River High can handle up to 225.
- The average cost per day of busing students is
  - $1.20 for each east-side student going to New High.
  - $2.00 for each east-side student going to River High.
  - $3.00 for each west-side student going to New High.
  - $1.50 for each west-side student going to River High.

continued ⬧

The problem facing the River City school board (and you) is to determine how many students to send to each school to minimize busing costs.

1. Write the constraints—that is, the equations and inequalities—that describe the situation. Use these variables.

   - N_e is the number of New High students who live on the east side of the river.
   - R_e is the number of River High students who live on the east side of the river.
   - N_w is the number of New High students who live on the west side of the river.
   - R_w is the number of River High students who live on the west side of the river.

2. Write the "cost of busing" expression (using the variables from Question 1).

3. List the combinations of constraints you will need to examine.

4. Solve the various combinations of equations you think you must look at.

5. Based on your solutions, write up your recommendation for the school board.

Fitting More Lines

In *Fitting a Line* your task was to find a function of the form $y = ax + b$ whose graph would go through the points $(1, 2)$ and $(-1, -6)$.

You began by finding several pairs of values for a and b for which the graph went through $(1, 2)$. You then looked for a relationship between a and b for those pairs.

You probably reached this conclusion.

> The line $y = ax + b$ goes through $(1, 2)$ if and only if the coefficients a and b fit the equation $a + b = 2$.

For example, the values $a = 7$, $b = -5$ fit this equation, and the line $y = 7x - 5$ goes through $(1, 2)$.

In the case of the point $(1, 2)$, the relationship between a and b is relatively simple. In this activity, you have a similar task, but you will use points for which the relationship may be somewhat harder to find.

A Line Through (3, 4) and (5, 1)

1. As with the other points you've considered, the diagram illustrates that there are infinitely many functions of the form $y = ax + b$ whose graphs go through the point $(3, 4)$.

   a. Find an equation involving a and b that guarantees that the line $y = ax + b$ goes through $(3, 4)$.

   b. Find a pair of numbers for a and b that fit your equation. Check whether the resulting line $y = ax + b$ actually does go through $(3, 4)$.

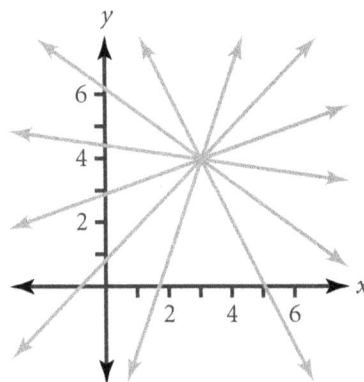

continued ◗

2. The diagram shows some of the lines through the point (5, 1).

   a. Find an equation involving a and b that guarantees that the line $y = ax + b$ goes through (5, 1).

   b. Find a pair of numbers for a and b that fit your equation. Check whether the resulting line $y = ax + b$ actually does go through (5, 1).

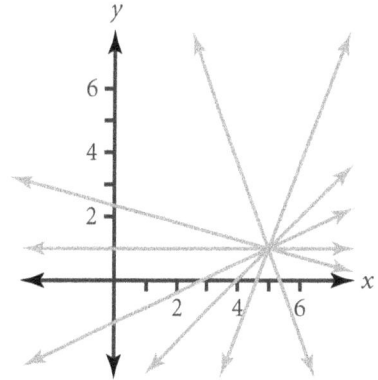

3. The diagrams for Questions 1 and 2 show some of the lines through (3, 4) and some of the lines through (5, 1). There is one line, not shown in either diagram, that goes through both points.

   a. Use the equations involving a and b from Questions 1a and 2a to find a line of the form $y = ax + b$ that goes through both (3, 4) and (5, 1).

   b. Verify that your answer is correct.

A Line Through (4, −1) and (−2, 7)

4. Use the method of Questions 1 to 3 to find a line of the form $y = ax + b$ that goes through both (4, −1) and (−2, 7).

Ages, Coins, and Fund-Raising

1. Paige is half as old as Quiana. Quiana is three years younger than Ryan. Ryan is nine years older than Paige.

   How old are Paige, Quiana, and Ryan?

2. Uncle Ralph has 18 coins in his pocket. Each coin is a quarter, a dime, or a nickel. The number of quarters and dimes combined is the same as the number of nickels. He has $2.10 worth of coins in his pocket.

   How many of each type of coin does he have?

3. Sonya, Dan, and Jesse were raising money for sports programs at their school. They had a pie sale, held a car wash, and sold raffle tickets.

   They charged $8 for each pie, charged $4 for each car wash, and sold the raffle tickets for $1 each. Altogether they collected $760.

   They spent $\frac{1}{2}$ hour on each car they washed and 2 hours making each pie they sold. These two activities required a total of 65 hours. They made twice as much money from raffle tickets as from car washes.

   How many pies did they sell? How many car washes did they do? How many raffle tickets did they sell?

Saved by the Matrices!

You've seen that solving a linear programming problem requires solving many systems of linear equations. And the unit problem has six variables! As you saw in the activity *Grind It Out,* even a four-variable system can require a huge amount of work.

Fortunately, your graphing calculator can assist you. Before you use the calculator, though, you will learn about matrices. A matrix is a notational shorthand for organizing a whole bunch of numbers.

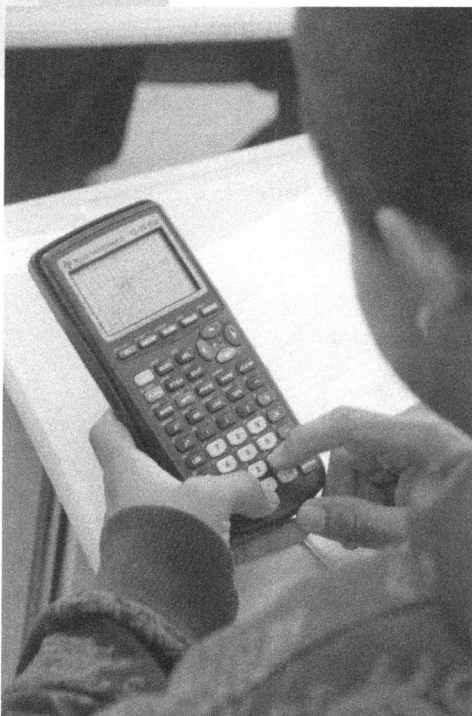

Giovanni Guzman uses a graphing calculator to address a system of linear equations with many variables.

Matrix Basics

A **matrix** is a rectangular array of numbers. Here are three examples

$$\begin{bmatrix} 1 & -8 & 3 & -1 & -4 \\ 5 & 3 & 4 & -1 & -6 \end{bmatrix} \qquad \begin{bmatrix} 1 & 1 & 1 \\ 2 & 1 & -1 \\ 3 & 2 & 2 \end{bmatrix} \qquad \begin{bmatrix} 6 & 17 & \frac{1}{8} & -368 \end{bmatrix}$$

The plural of matrix is *matrices.* Matrices are often written using parentheses instead of brackets. Either method is correct.

The matrix $\begin{bmatrix} 1 & -8 & 3 & -1 & -4 \\ 5 & 3 & 4 & -1 & -6 \end{bmatrix}$ has two rows and five columns, so it is called a *2-by-5 matrix.* A row is horizontal; a column is vertical. The pair of numbers -4 and -6 form one column of this matrix.

The numbers 2 and 5 are called the **dimensions of the matrix.** The phrase "2-by-5" is usually represented in writing as 2×5, just as we refer to a rectangle that is 3 inches wide and 4 inches long as a "3×4 rectangle." The first dimension of a matrix tells how many rows the matrix has. The second dimension tells how many columns it has.

The matrix $\begin{bmatrix} 1 & 1 & 1 \\ 2 & 1 & -1 \\ 3 & 2 & 2 \end{bmatrix}$ has three rows and three columns, so it is a 3×3 matrix. The matrix $\begin{bmatrix} 6 & 17 & \frac{1}{8} & -368 \end{bmatrix}$ has one row and four columns, so it is a 1×4 matrix.

A matrix with the same number of rows as columns, such as a 3×3 matrix, is called a **square matrix.** A matrix with only one row is often called a **row vector,** and a matrix with only one column is often called a **column vector.** Each individual number in a matrix is called an **entry.**

Inventing an Algebra

Matrix Basics defines what a matrix is and introduces you to the standard notation and terminology of matrices. There is also an *algebra* of matrices, which means there are rules for adding and multiplying them.

This activity will help you discover some of those rules by using matrices in meaningful contexts.

1. A matrix could be used to keep track of students' points in a class. Each row could stand for a different student, such as Ana, Ben, Cass, and Devon. The first column might be for homework, the second for oral reports, and the third for POWs.

   Suppose this table represents the results for the first grading period.

	Homework	Reports	POWs
Ana	18	54	30
Ben	35	23	52
Cass	46	15	60
Devon	60	60	60

   A matrix representation of this information might look like this.

$$\begin{bmatrix} 18 & 54 & 30 \\ 35 & 23 & 52 \\ 46 & 15 & 60 \\ 60 & 60 & 60 \end{bmatrix}$$

continued ▶

Here, in table form, are the students' points in each category for the second grading period.

	Homework	Reports	POWs
Ana	10	60	0
Ben	52	35	58
Cass	42	20	48
Devon	60	60	60

a. Write these second-grading-period scores in a matrix.

b. Compute each student's total points *in each assignment category* for the two grading periods combined. Write the totals in matrix form.

c. Congratulations! If you completed part b, you have added two matrices. Based on your work, write an equation showing two matrices being added to give your matrix from part b.

2. The Woos' bakery is open six days a week.

Last week, for chocolate chip cookies, the Woos sold 30 dozen cookies on Monday, 25 dozen on Tuesday, 27 dozen on Wednesday, 23 dozen on Thursday, 38 dozen on Friday, and 52 dozen on Saturday.

For plain cookies, they sold 30 dozen on Monday, 28 dozen on Tuesday, 20 dozen on Wednesday, 25 dozen on Thursday, 35 dozen on Friday, and 45 dozen on Saturday.

For iced cookies, they sold 45 dozen on Monday, 32 dozen on Tuesday, 40 dozen on Wednesday, 38 dozen on Thursday, 48 dozen on Friday, and 70 dozen on Saturday.

a. Use a matrix to represent the Woos' sales. Let each row represent a different kind of cookie and each column represent a different day of the week.

b. Make up sales numbers for the Woos for a second week. Show your sales in a matrix similar to that for part a.

c. Add the Woos' sales for the two weeks. Show the totals in a matrix similar to those for parts a and b.

d. Write the matrix addition equation that corresponds to your work.

3. Which of the matrix sums shown here do you think make sense? Find those sums, and explain why you think the other sums don't make sense.

a. $\begin{bmatrix} 1 & 5 & 0 & -6 \\ 2 & -2 & 4 & 1 \\ 0 & 1 & -3 & 1 \end{bmatrix} + \begin{bmatrix} 8 & -4 & 0 & 3 \\ 3 & 2 & 4 & 5 \\ 1 & -3 & 3 & 6 \end{bmatrix}$

b. $\begin{bmatrix} 1 & 5 & 0 & -6 \\ 2 & -2 & 4 & 1 \end{bmatrix} + \begin{bmatrix} 8 & -4 & 0 \\ 3 & 3 & 2 \\ 7 & 5 & 1 \end{bmatrix}$

c. $[-3 \ \ 7 \ \ 9] + [7 \ \ -3 \ \ 5]$

d. $[5 \ \ -4 \ \ 2 \ \ 1] + \begin{bmatrix} 4 \\ -2 \\ 7 \\ 1 \end{bmatrix}$

4. What has to be true of two matrices for it to make sense to add them together?

5. Describe a rule for adding any matrices that fit your condition from Question 4.

Fitting Quadratics

In *Fitting a Line* and *Fitting More Lines,* your task was to find a linear function through a particular pair of points. That is, you were asked to find coefficients a and b so that the line $y = ax + b$ would go through the given points.

In this activity, you move from linear functions to another important category: the family of **quadratic** functions. A quadratic function is often written in the form $y = ax^2 + bx + c$. In this equation, the coefficients a, b, and c can be any three numbers, not necessarily different, except that a cannot be zero. If $a = 0$, the function is linear. (*Reminder:* The graph of a quadratic equation is a shape called a *parabola.*)

As you might suspect, because a quadratic function has three coefficients, it generally takes three points on its graph to determine the function. In Questions 1 to 3, you will consider all the parabolas going through a particular point. Question 4 asks you to put the information from those questions together.

1. The diagram shows some of the parabolas that go through the point $(1, 6)$.

   a. Verify that the parabola defined by the equation $y = 3x^2 + 2x + 1$ goes through $(1, 6)$.

   b. Find at least two other possible choices of values for a, b, and c so that the parabola $y = ax^2 + bx + c$ goes through $(1, 6)$.

   c. Find an equation involving a, b, and c that holds true for every parabola $y = ax^2 + bx + c$ that goes through $(1, 6)$.

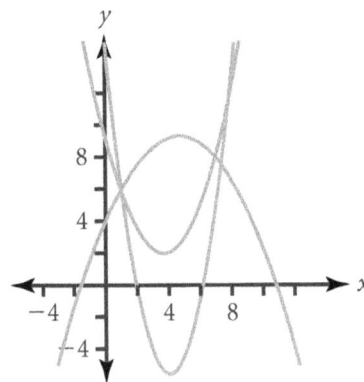

2. Now consider the set of parabolas that go through the point $(3, 8)$. Find an equation involving a, b, and c that holds true for every parabola $y = ax^2 + bx + c$ that goes through $(3, 8)$.

3. Find an equation involving a, b, and c that holds true for every parabola $y = ax^2 + bx + c$ that goes through $(5, 2)$.

continued

4. a. Use the equations you found in Questions 1, 2, and 3 to find a parabola that goes through all three points: $(1, 6)$, $(3, 8)$, and $(5, 2)$. That is, find values for a, b, and c so that the parabola $y = ax^2 + bx + c$ goes through all three points.

   b. Is your solution to Question 4a unique? Explain.

Flying Matrices

You've seen that there is a natural way to add matrices. Namely, if the matrices have the same dimensions, you simply add the corresponding entries. (If the matrices have different dimensions, you can't add them.)

Defining multiplication of matrices is harder, and the definition is somewhat arbitrary. The arithmetic in these questions will be the basis for deciding what might make sense when defining how matrices are multiplied.

1. In *Heavy Flying,* you were given these facts about the materials that Lindsay transports.

   • Charley's Chicken Feed packages its product in containers that weigh 40 pounds and are 2 cubic feet in volume.

   • Careful Calculators packages its materials in boxes that weigh 50 pounds and are 3 cubic feet in volume.

   Organize this information into a matrix. Label the rows and columns to show what the numbers represent.

2. Suppose that on Monday, Lindsay transports 500 containers of chicken feed and 200 boxes of calculators. Put those facts into a matrix.

3. Use the information in your two matrices to find the total weight and the total volume Lindsay transported on Monday. Put those two answers into a matrix.

4. Explain how you used the information in the matrices you created in Questions 1 and 2 to calculate the numbers for Question 3.

5. Suppose that on Tuesday, Lindsay transports 400 containers of chicken feed and 300 boxes of calculators. Combine this information with the data from Question 2 to form a 2×2 matrix showing exactly what she carried on Monday and on Tuesday.

continued ▶

6. Combine the information in your answers to Questions 1 and 5 to find the total weight and the total volume transported on Monday and on Tuesday. Find separate totals for each day. Put all of that information into a matrix.

7. Explain how you calculated the numbers for your matrix in Question 6.

Matrices in the Oven

The Woos' bakery provides another example for multiplying matrices. The questions in this activity are similar to those in *Flying Matrices*. Pay careful attention to the arithmetic they involve.

For this activity, you can ignore the constraints from the *More Cookies* problem. The facts about the ingredients remain the same.

- One dozen plain cookies requires 1 pound of cookie dough (and no icing or chocolate chips).
- One dozen iced cookies requires 0.7 pound of cookie dough and 0.4 pound of icing (and no chocolate chips).
- One dozen chocolate chip cookies requires 0.9 pound of cookie dough and 0.15 pound of chocolate chips (and no icing).

1. Put all of this information into a matrix. Be sure to label your rows and columns.

2. Suppose that on Wednesday, the Woos made 30 dozen plain cookies, 45 dozen iced cookies, and 30 dozen chocolate chip cookies. On Thursday, they made 28 dozen plain cookies, 32 dozen iced cookies, and 25 dozen chocolate chip cookies.

   Put all of this information into a matrix.

3. Combine the information in your answers to Questions 1 and 2 to create a matrix that shows the total amount *of each ingredient* used on Wednesday and on Thursday.

4. Describe how you calculated the numbers for your matrix in Question 3.

Fresh Ingredients

The Woos buy their ingredients fresh each day. They shop at two markets, and the prices are slightly different at each.

It takes too much time for them to go to both markets. Each day they have to decide where to shop, depending on what they are baking that day.

The Woos' baking plan for Wednesday and Thursday is the one described in *Matrices in the Oven*. It can be represented by this matrix.

$$
\begin{array}{c}
 \\
\\
\text{Wed} \\
\text{Thurs}
\end{array}
\begin{array}{ccc}
 & & \text{choc} \\
\text{plain} & \text{iced} & \text{chip} \\
\left[\begin{array}{ccc} 30 & 45 & 30 \\ 28 & 32 & 25 \end{array}\right]
\end{array}
$$

The first 30 in the matrix, for example, means they will make 30 dozen plain cookies on Wednesday.

You also need the information in this ingredient matrix.

$$
\begin{array}{c}
 \\
\\
\text{plain} \\
\text{iced} \\
\text{choc chip}
\end{array}
\begin{array}{ccc}
 & & \text{choc} \\
\text{dough} & \text{icing} & \text{chips} \\
\left[\begin{array}{ccc} 1 & 0 & 0 \\ 0.7 & 0.4 & 0 \\ 0.9 & 0 & 0.15 \end{array}\right]
\end{array}
$$

The entry 0.7, for example, means the Woos need 0.7 pound of cookie dough for each dozen iced cookies.

Now, if the Woos shop at Farmer's Market, their ingredient costs are

- dough: 30¢ per pound
- icing: 20¢ per pound
- chocolate chips: 32¢ per pound

If they shop at Downtown Grocery, their costs are

- dough: 29¢ per pound
- icing: 28¢ per pound
- chocolate chips: 22¢ per pound

continued ▶

The prices at the stores don't change from Wednesday to Thursday. The Woos want to know what it would cost to do all of their Wednesday shopping at Farmer's Market or all of it at Downtown Grocery. They want similar information for Thursday.

1. Put all the pricing information for the two markets into a single matrix. Set up your matrix in a way that will allow you to multiply matrices to get the cost information the Woos need.

2. Use matrices to help the Woos decide which market to shop at on Wednesday and which to shop at on Thursday.

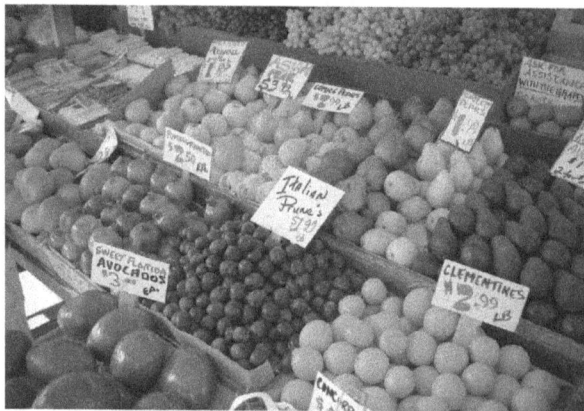

Calculators to the Rescue

You've seen that multiplying matrices can involve a lot of cumbersome arithmetic. Even without matrices, you would still have to do the same arithmetic to solve problems like those in *Fresh Ingredients*. And this type of computation is done in businesses every day.

Fortunately you can avoid all this arithmetic by entering the information as matrices into a graphing calculator and letting the calculator do the work.

1. The Woos' baking plan is given by this matrix.

	plain	iced	choc chip
Wed	30	45	30
Thurs	28	32	25

   Give this matrix a name and enter it into your calculator.

2. The amount of each ingredient for each type of cookie is given by this matix.

	dough	icing	choc chips
plain	1	0	0
iced	0.7	0.4	0
choc chip	0.9	0	0.15

   Enter this matrix into your calculator. Give it a different name from the one you used in Question 1.

3. The costs of the ingredients at each market are given by this matrix

	Frmr Mkt	Dntn Groc
dough	30	29
icing	20	28
choc chips	32	22

   Enter this matrix into your calculator, giving it a third name.

continued ⟩

4. In *Fresh Ingredients,* you found a matrix showing how much all the ingredients would cost the Woos at each market on each day—Wednesday and Thursday. Check that result using your calculator.

Make It Simple

1. Write instructions explaining how to multiply two 3 × 3 matrices (using pencil and paper, not a calculator). Make your instructions very clear.

2. Give your instructions to someone who has not learned how to multiply matrices. See whether he or she can use your instructions to multiply two 3 × 3 matrices that are different from those in your instructions.

3. Describe any difficulties the person may have had in following your instructions. Indicate what may have been wrong with the instructions.

4. Revise your instructions as necessary.

Back and Forth

You have seen that a single linear equation can be written as a matrix equation. More specifically, the equation can be represented as a statement that the product of two matrices is a certain number. (Technically, the product is a 1×1 matrix, but we usually think of it as just a number.)

For example, the equation

$$5x + 3y + 7z = 10$$

states that the product of the row matrix $[5 \ 3 \ 7]$ and the

column matrix $\begin{bmatrix} x \\ y \\ z \end{bmatrix}$ is equal to 10. So the original equation

$5x + 3y + 7z = 10$ says the same as the matrix equation

$$[5 \ 3 \ 7] = \begin{bmatrix} x \\ y \\ z \end{bmatrix}$$

Now consider this system of linear equations.

$$5x + 3y + 7z = 10$$
$$2x + y - z = 1$$
$$3x + 2y + z = 4$$

1. Develop a way to write this system of equations as a single matrix equation. Think of the numbers on the right of the equal signs as forming a column vector.

2. Now turn the process around by writing this matrix equation as a pair of linear equations.

$$\begin{bmatrix} 3 & -2 \\ 1 & -6 \end{bmatrix} \begin{bmatrix} u \\ v \end{bmatrix} = \begin{bmatrix} 2 \\ 1 \end{bmatrix}$$

continued ⬧

For the rest of these questions, either turn the system of linear equations into a single matrix equation or write the matrix equation as a system of linear equations. You do not have to solve any of these equations.

3. $3a - 2b = 7$
   $-4a + 5b = 9$

4. $\begin{bmatrix} 1 & 0 \\ -2 & 3 \end{bmatrix} \begin{bmatrix} r \\ s \end{bmatrix} = \begin{bmatrix} 2 \\ -3 \end{bmatrix}$

5. $3c = 4$
   $-c + 2d = -5$

6. $\begin{bmatrix} 1 & -1 & 2 \\ 0 & 3 & 0 \\ 3 & -3 & 1 \end{bmatrix} \begin{bmatrix} e \\ f \\ g \end{bmatrix} = \begin{bmatrix} 2 \\ -4 \\ 1 \end{bmatrix}$

7. $3x + 2y - z = 4$
   $3y + 2z = 0$
   $5x - 3z = 1$

A system of linear equations can be expressed as a single matrix equation, which will look like

$$[A][X] = [B]$$

In this equation, [A] is the **coefficient matrix** of the system of linear equations and [B] is a column matrix made up of the numbers on the right of the equal signs in the linear equations. (This column matrix is sometimes called the **constant term matrix.**) The matrix [X] is a column matrix made up of the individual variables in the system of linear equations.

For example, consider this pair of equations.

$$3a - 2b = 7$$
$$-4a + 5b = 9$$

The coefficient matrix [A] is $\begin{bmatrix} 3 & -2 \\ -4 & 5 \end{bmatrix}$, the constant term matrix [B]

is $\begin{bmatrix} 7 \\ 9 \end{bmatrix}$, and [X] is the column matrix $\begin{bmatrix} a \\ b \end{bmatrix}$.

Solving the pair of equations is the same as finding values for a and b so that

$$\begin{bmatrix} 3 & -2 \\ -4 & 5 \end{bmatrix} \begin{bmatrix} a \\ b \end{bmatrix} = \begin{bmatrix} 7 \\ 9 \end{bmatrix}$$

Any system of linear equations can be expressed as a matrix equation. Similarly, an appropriate matrix equation [A][X] = [B], where [X] is a matrix of variables, can be interpreted as a system of linear equations.

Solving the Simplest

1. Solve these matrix equations. You will probably want to turn each one into a system of linear equations.

   a. $\begin{bmatrix} 1 & 2 \\ 3 & 4 \end{bmatrix} \begin{bmatrix} w \\ z \end{bmatrix} = \begin{bmatrix} 6 \\ 16 \end{bmatrix}$

   b. $\begin{bmatrix} 1 & 3 \\ 0 & 2 \end{bmatrix} \begin{bmatrix} r \\ s \end{bmatrix} = \begin{bmatrix} 7 \\ 2 \end{bmatrix}$

   c. $\begin{bmatrix} 2 & 0 \\ 0 & 3 \end{bmatrix} \begin{bmatrix} u \\ v \end{bmatrix} = \begin{bmatrix} 8 \\ 15 \end{bmatrix}$

2. Compare the three matrix equations in Question 1. Which coefficient matrix led to the most easily solved system of linear equations?

3. Find the 2 × 2 coefficient matrix that will give the system of linear equations that is the absolute easiest to solve. Explain your choice.

Things We Take for Granted

We often take for granted all sorts of details about mathematics. For instance, properties like $8 \cdot 5 = 5 \cdot 8$ seem so natural and obvious that we may not even think about them.

In this activity, you will consider two properties that are true for numbers and explore whether similar properties hold true for 2×2 matrices.

1. The Commutative Property of Multiplication

The **commutative** property of multiplication for numbers is illustrated by the example $8 \cdot 5 = 5 \cdot 8$. More generally, according to this property, when you change the order of the quantities you are multiplying, you get the same answer. We can express this property in symbols as $ab = ba$.

Does a similar property hold true for multiplication of 2×2 matrices? See if you can find a counterexample.

2. The Associative Property of Multiplication

The **associative** property of multiplication for numbers applies when three numbers are multiplied together. This property says that regrouping the factors to change which multiplication is performed first does not change the result. An example is $(3 \cdot 4) \cdot 5 = 3 \cdot (4 \cdot 5)$. The left side is $12 \cdot 5$, and the right side is $3 \cdot 20$, and both sides equal 60.

Does a similar property hold true for multiplication of 2×2 matrices? See if you can find a counterexample.

Finding an Inverse

You have seen that you can solve a matrix equation like

$$\begin{bmatrix} 1 & 2 \\ 3 & 5 \end{bmatrix} \begin{bmatrix} w \\ z \end{bmatrix} = \begin{bmatrix} 2 \\ 1 \end{bmatrix}$$

by first finding a matrix C that fits the equation

$$[C] \begin{bmatrix} 1 & 2 \\ 3 & 5 \end{bmatrix} = \begin{bmatrix} 1 & 0 \\ 0 & 1 \end{bmatrix}$$

If there is such a matrix [C], it is called the **multiplicative inverse** of $\begin{bmatrix} 1 & 2 \\ 3 & 5 \end{bmatrix}$.

To look for this inverse, suppose that [C] is the matrix $\begin{bmatrix} r & s \\ t & u \end{bmatrix}$. The equation defining [C] then becomes

$$\begin{bmatrix} r & s \\ t & u \end{bmatrix} \begin{bmatrix} 1 & 2 \\ 3 & 5 \end{bmatrix} = \begin{bmatrix} 1 & 0 \\ 0 & 1 \end{bmatrix}$$

1. Multiply out the product $\begin{bmatrix} r & s \\ t & u \end{bmatrix} \begin{bmatrix} 1 & 2 \\ 3 & 5 \end{bmatrix}$ to get a matrix with entries in terms of r, s, t, and u.

2. Use the fact that the product in Question 1 is equal to $\begin{bmatrix} 1 & 0 \\ 0 & 1 \end{bmatrix}$ to get a system of linear equations involving r, s, t, and u.

3. Solve the equations from Question 2 to find the values of r, s, t, and u.

4. a. Use your results from Question 3 to write the matrix [C].
   b. Check your work by finding the product $[C] \begin{bmatrix} 1 & 2 \\ 3 & 5 \end{bmatrix}$.

Inverses and Equations

1. You know that some systems of linear equations have unique solutions, some have no solutions, and some have infinitely many solutions. For each of these systems, find the unique solution or explain why there is not a unique solution.

   a. $x + 2y = 4$
   $3x + 4y = 8$

   b. $x + 2y = 2$
   $3x + 4y = 1$

   c. $x + 2y = 4$
   $2x + 4y = 8$

   d. $x + 2y = 2$
   $2x + 4y = 1$

2. The matrices in parts a and b here are each the coefficient matrix for two of the systems of equations in Question 1. For each matrix, either find the **inverse** or explain why there is no inverse.

   a. $\begin{bmatrix} 1 & 2 \\ 3 & 4 \end{bmatrix}$

   b. $\begin{bmatrix} 1 & 2 \\ 2 & 4 \end{bmatrix}$

3. Explain how your results in Question 2 are related to those in Question 1.

4. In general, which 2×2 matrices have inverses and which do not? What is the connection between a system of equations and whether a matrix has an inverse?

Calculators Again

You know that you can use a graphing calculator to multiply matrices. Graphing calculators have another wonderful use. They can calculate the multiplicative inverse of a matrix when the matrix is invertible (as long as the matrix dimensions are not too large for the calculator's capacity). And this is easy once you have entered a matrix into the calculator's memory.

Try to solve the following linear systems without doing any arithmetic, letting the calculator do all the hard work for you. Check your solutions, at least for the first system, to make sure you are doing the process correctly.

1. $5d + 2e = 11$
   $d + e = 4$

2. $2r + 3s - t = 3$
   $r - 2s + 4t = 2$
   $4r - s + 7t = 8$

3. $4w + x + 2y - 3z = -16$
   $-3w + x - y + 4z = 20$
   $-w + 2x + 5y + z = -4$
   $5w + 4x + 3y - z = -10$

4. $a + b + c + d + e + f = 30$
   $2a + 3b - 6c + 4d - e + f = 8$
   $5a + 4b + 3c - d + 5e - 2f = 34$
   $2a - 3b + 8c - 6d + e + 4f = 38$
   $6a + 2b + 7c - 5d - 3e - 2f = -42$
   $-5a + 8b - 5c + 3d - 9e + 4f = -18$

Fitting Mia's Birdhouses

Mia and her friends spent the semester building birdhouses, and now
they are painting them. (You may remember Mia and her birdhouses
from the Year 1 unit *The Pit and the Pendulum.*) After one hour, they
had painted two birdhouses. After three hours, they had painted six
birdhouses. And after five hours, they had painted nine birdhouses.

1. Plot the data about birdhouse painting. Use *number of hours* for the
   x-axis and *number of birdhouses* for the *y*-axis.

2. Explain why there is no linear function that fits the data perfectly.

3. Find a quadratic function that does fit the data perfectly. That
   is, find values for *a*, *b*, and *c* so that all three data points from the
   birdhouse situation fit the equation $y = ax^2 + bx + c$.

4. What do you think about the usefulness of the function from
   Question 3 in this situation?

Solving *Meadows or Malls?*

Congratulations! You are now ready to solve the unit problem.

With so many constraints and so many variables, it won't be easy—but careful work and a few shortcuts will help you complete the task.

You will also be preparing your unit portfolio in this final portion of the unit. Because this unit involves so many ideas, your portfolio work will be spread out over several activities.

Jon Honn and Rodolfo Contreras make sure the list of combinations each developed is complete before they begin to solve the unit problem.

Getting Ready for *Meadows or Malls?*

A while back (it may seem like a year ago!), you started working on the *Meadows or Malls?* problem. You now have all the tools you need to solve that problem, although it still will take some work to answer the main question in that problem.

As you have seen, the *Meadows or Malls?* problem has 12 constraints, of which four are equations and eight are inequalities. Here are those constraints.

$$\text{I} \qquad G_R + G_D = 300$$

$$\text{II} \qquad A_R + A_D = 100$$

$$\text{III} \qquad M_R + M_D = 150$$

$$\text{IV} \qquad G_D + A_D + M_D \geq 300$$

$$\text{V} \qquad A_R + M_R \leq 200$$

$$\text{VI} \qquad A_R + G_D = 100$$

$$\text{VII} \qquad G_R \geq 0$$

$$\text{VIII} \qquad A_R \geq 0$$

$$\text{IX} \qquad M_R \geq 0$$

$$\text{X} \qquad G_D \geq 0$$

$$\text{XI} \qquad A_D \geq 0$$

$$\text{XII} \qquad M_D \geq 0$$

These constraints give you a lot of combinations to check to find corner points of the feasible region. Your task now is to list all the combinations you need to check. Refer to the constraints by number in your list of combinations.

continued ▶

As you work, keep these things in mind.

- The four equations must be part of every combination.

- Look for ways to reduce your list (without actually solving any of the systems). For instance, it is possible to prove from the constraints that certain variables can't be zero. Once you do that, you can eliminate some of constraints VII through XII from consideration.

Meadows or Malls? Revisited

River City has three parcels of land to work with.

- 300 acres of farmland willed to the city by Mr. Goodfellow
- 100 acres from the U.S. government from a closed army base
- 150 acres of land formerly leased for mining

The city needs to decide how much of each parcel to use for recreation and how much to use for development, based on several constraints.

Your Task

Remember that your group is a consulting team that the city manager has come to for help. Not only should you give her an answer, but you should also try to convince her that yours is the best possible answer. You'd like her to use your group in the future when the city needs help again.

Prepare a group report for the city manager. Your report should include three parts.

- An answer to the city's dilemma
- An explanation that will convince the city manager that your solution will cost the least
- Any graphs, charts, equations, or diagrams that are needed to support your explanation

You will probably want to review what you already know about the problem based on your notes and the previous activities.

Beginning Portfolios—Part I

This unit involves several closely related ideas.

- Graphing linear equations in three variables
- Solving systems of linear equations in three variables
- Finding intersections of planes in 3-space

1. Summarize how these ideas are related. In particular, focus on these two questions and how they are connected.

   - What are the possible results from solving a system of three linear equations in three variables?

   - What are the possible results of the intersection of three planes in 3-space?

2. Select activities from the unit that were important in developing your understanding of the ideas you discussed in Question 1. Explain why you made the selections you did.

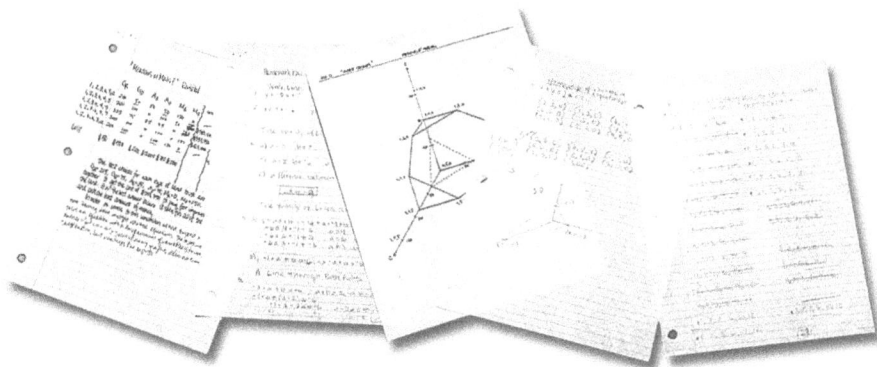

Beginning Portfolios—Part II

In this unit, you have used matrices both to represent information and to solve linear equations.

1. Summarize what you have learned about matrices. Answer these questions in your summary.

   • What is a matrix?

   • How are matrices used to solve linear equations?

2. Select activities from the unit that were important in developing your understanding of the ideas you discussed in Question 1. Explain why you made the selections you did.

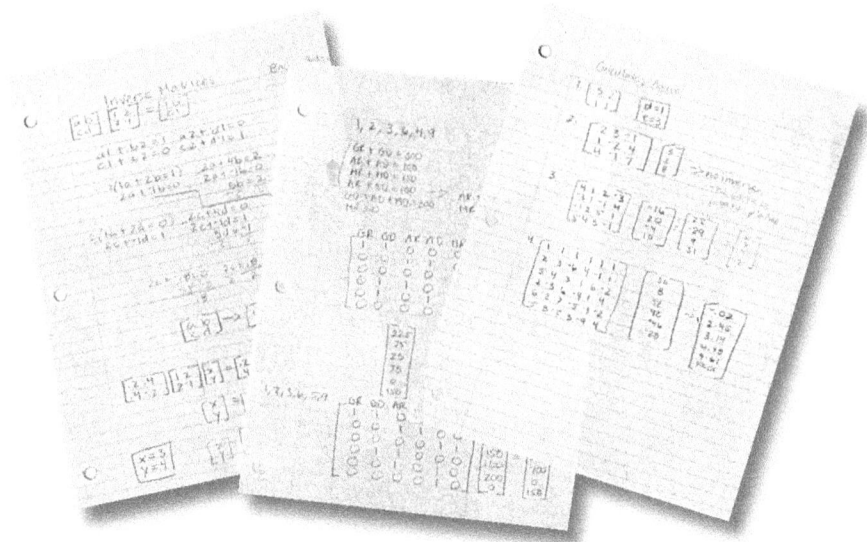

Meadows or Malls? Portfolio

Now that *Meadows or Malls?* is completed, it is time to assemble your portfolio for the unit. This process has three steps.

- Write a cover letter that summarizes the unit.
- Choose papers to include from your work in the unit.
- Discuss your personal mathematical growth during the unit.

Cover Letter

Look back over *Meadows or Malls?* and describe the central problem of the unit and the key mathematical ideas. Your description should give an overview of how the key ideas were developed and how they were used to solve the central problem.

In compiling your portfolio, you will select some activities you think were important in developing the unit's key ideas. Your cover letter should include an explanation of why you selected each item. For ideas discussed in *Beginning Portfolios—Part I* and *Beginning Portfolios— Part II,* you can simply refer to your work on those activities.

continued

Selecting Papers

Your portfolio for *Meadows or Malls?* should contain these items.

- *Meadows or Malls? Revisited*
- *Beginning Portfolios—Part I* and *Beginning Portfolios—Part II*

  Include these two earlier portfolio assignments and the activities you discussed in them as part of your cover letter.
- *Just the Plane Facts*
- An activity on solving systems of linear equations that does not use matrices
- An activity in which you learned concepts that allowed you to solve linear programming problems in more than two variables
- A Problem of the Week

  Select one of the POWs you completed in this unit: *That's Entertainment!, SubDivvy,* or *Crack the Code.*

Personal Growth

Your cover letter for *Meadows or Malls?* describes how the mathematical ideas develop in the unit. In addition, write about your personal development during this unit. You may want to address this question.

  How did your experiences in the Year 2 unit Cookies *affect your work in* Meadows or Malls?

Include any thoughts about your experiences that you wish to share with a reader of your portfolio.

SUPPLEMENTAL ACTIVITIES

The supplemental activities in *Meadows or Malls?* touch on algebraic and geometric ideas, as well as principles of linear programming. Here are some examples.

- *Embellishing the Plane Facts* asks you to find equations to illustrate some of your conclusions from *Just the Plane Facts*.

- *Producing Programming Problems—More Variables* asks you to create a linear programming problem involving at least four variables.

- *Determining the Determinant* asks you to investigate the concept of a determinant, which is a useful tool in deciding whether a matrix is invertible.

How Many Regions?

Any line in the coordinate plane will divide the plane into two regions, one on either side of the line. (The line itself is not considered a region.) Each of these regions can be represented by an inequality.

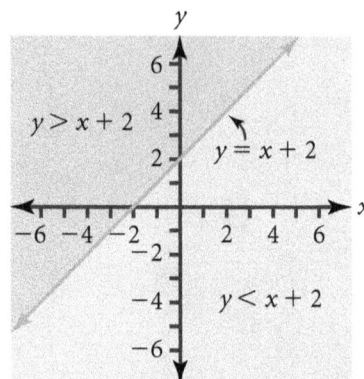

For example, the diagram shows the line defined by the equation $y = x + 2$. The line itself contains all the points that fit the equation, such as $(-3, -1)$ and $(2, 4)$.

The darkly shaded region above the line consists of points that fit the inequality $y > x + 2$, such as $(0, 3)$ and $(-4, 1)$. The lightly shaded region below the line consists of points that fit the inequality $y < x + 2$, such as $(3, 4)$ and $(-2, -2)$.

Your task is to investigate what happens if you start with more than one line. Specifically, how many regions might you get, and how can you describe the different regions using inequalities?

1. Begin with the case of two lines.

   a. What are the possibilities for the number of regions you get when you draw two lines in the plane?

   b. For each possibility in part a, give a specific pair of equations you could use for the lines, and describe how to represent each resulting region using inequalities. Some regions may require more than one inequality.

   c. Explain why you think your answer to part a includes all the possibilities.

continued ▶

2. Do the same thing for the case of three lines and then for the case of four lines. Give specific examples for each possibility you describe, and explain why you think you have covered all the possibilities.

The Eternal Triangle

Krys, Greg, and Juranso are creating a musical love story based on the life of Pythagoras. Krys will design the sets, Greg will compose the music, and Juranso will write the script. Greg and Juranso will both perform in the musical.

The three have agreed that the play will be a mixture of straight acting scenes and musical scenes. They need to decide how many of each type of scene to include.

Krys, Greg, and Juranso have a limited amount of time for the project. Krys can design at most 16 sets. There will be one set per scene, no matter what kind of scene it is. Greg and Juranso can each spend at most 36 hours on the play. Greg figures it will take him an hour for each acting scene but, because he has to write the music, three hours for each musical scene. For Juranso, it's the opposite. Because he's developing the script, he figures the acting scenes will take him three hours each and the musical scenes will require only one hour each.

Of course, the trio wants to maximize the number of people who will attend their performance. Juranso does a lot of acting, and they figure Greg's musical scenes will be a special draw. They estimate that 20 people will attend for each musical scene they include and 10 people will attend for each acting scene they include.

How many of each type of scene should they include in their play to maximize the number of attendees?

Note: You can interpret an answer with fractions as representing a scene that is partly music and partly straight acting, so don't limit yourself to whole numbers.

The Jewelry Business

Rebecca, Noel, and Keenzia have decided to start a jewelry business. They will buy hand-crafted jewelry in bulk at a discount and sell it at craft fairs and art shows.

To purchase jewelry at the special bulk price, they need to spend at least $3,000. The earrings will cost them $5 per pair. The necklaces will cost them $9 apiece.

They have different opinions about what portion of their purchase should be for earrings and what portion should be for necklaces. At this point, all they've decided is that the number of necklaces they buy should be at least half but at most three times the number of pairs of earrings they buy.

Aside from the time spent planning and purchasing, the partners estimate it will take an average of 15 minutes to sell each pair of earrings and 20 minutes to sell each necklace. Together, the group can devote a total of up to 240 hours to the endeavor.

After all their costs are considered, they make $4 profit on each necklace and $3 on each pair of earrings.

Assuming they sell everything they buy (and they will, because they are shrewd businesspeople), how many necklaces and how many pairs of earrings should they buy to maximize their profit?

Special Planes

You know that the graph of a linear equation in three variables
is a plane. Now you will find equations for planes that fit some
specific conditions.

1. a. Find a linear equation whose graph is a plane
      that includes the z-axis. (Consider what's true
      about the coordinates of all points on the
      z-axis.)

   b. Find a linear equation whose graph differs
      from the graph of your equation from part a
      but still includes the z-axis.

2. a. Find a linear equation whose graph includes
      the points $(0, 0, 0)$ and $(1, 1, 0)$.

   b. Find a linear equation whose graph differs
      from the graph of your equation from part a
      but still includes the points $(0, 0, 0)$ and $(1, 1, 0)$.

3. There are many planes whose graph includes both $(0, 0, 0)$
   and $(1, 1, 0)$.

   a. In what kind of set do all these planes intersect?

   b. What points in addition to $(0, 0, 0)$ and $(1, 1, 0)$ must be in the
      set where all these planes intersect?

4. a. Find a linear equation whose graph includes the point $(2, -1, 4)$.

   b. Find a linear equation whose graph includes both $(2, -1, 4)$
      and $(1, 1, 0)$.

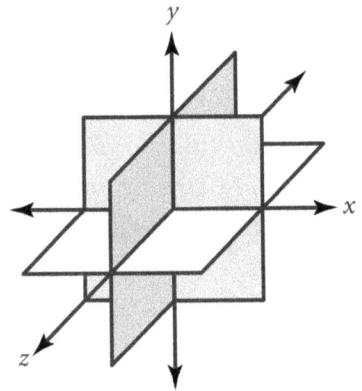

Embellishing the Plane Facts

In the activity *Just the Plane Facts,* you described the possible ways in which two or more planes might intersect. You explained each case using a diagram or model.

Now you will describe each case by giving equations of specific planes that create the given type of intersection.

1. First consider the case of two planes.

   a. Describe the different ways in which two planes can intersect. (This is Question 2 of *Just the Plane Facts.*)

   b. For each kind of intersection, give the equations of two planes that intersect in that way. Do not use the coordinate planes for either of the two planes.

   c. How can you tell by looking at the equations for two planes how the planes will intersect?

2. Now consider the case of three planes.

   a. Describe the different ways in which three planes can intersect.

   b. For each kind of intersection, give the equations of three planes that intersect in that way. Do not use the coordinate planes for any of the three planes.

A Linear Medley

Linear equations and systems of equations arise in many types of situations. Examine the situations given here. Express each situation in terms of linear equations, and then try to answer the questions by solving the equations.

If a problem does not have a solution, or does not give enough information to answer the questions, propose a way to change the problem so that it will have a unique solution.

1. Hound Dog Busline charges passengers $50 to ride to Nashville plus $5 for each piece of luggage. King Busline charges $70 for the trip and only $3 per piece of luggage.

   a. Write an expression for how much Hound Dog Busline charges if a passenger brings b pieces of luggage.

   b. Write an expression for how much King Busline charges if a passenger brings b pieces of luggage.

   c. For what number of pieces of luggage carried by a passenger do the two companies charge the same amount? What is the charge?

2. The length of a certain rectangle is 9 times its width. The perimeter of the rectangle is 200 meters. What are the dimensions of the rectangle?

3. When Nicole sold 8 pounds of tomatoes and 5 pounds of apples, she made $12.10. When she sold 12 pounds of tomatoes and 9 pounds of apples, she made $20.10. How much does Nicole charge per pound for tomatoes and for apples?

4. Six hundred people attended the city basketball championship. Two types of tickets were sold: adult tickets and student tickets. A total of $2,200 was collected at the gate. How much did each type of ticket cost?

continued ▶

5. Rancher Gonzales hired some workers to put up a fence around her ranch. On the first day, she hired four experienced carpenters and six apprentices, and they completed 420 feet of fencing. The next day, she hired five experienced carpenters and three apprentices, and they completed 390 feet of fencing.

For simplicity, assume the workers all work independently, so the amount of fencing completed is simply the sum of the amounts completed by each individual. Also assume that each experienced carpenter does the same amount of work and that each apprentice does the same amount of work.

Based on these assumptions, how much fencing does each experienced carpenter finish per day? How much fencing does each apprentice finish per day?

The General Two-Variable System

You have seen that every linear equation in one variable can be solved using basic ideas about equivalent equations.

The standard linear equation has the form $ax + b = c$. You can get from $ax + b = c$ to a solution for x in terms of the other variables using this sequence of equations.

$$ax + b = c$$
$$ax = c - b$$
$$x = \frac{c - b}{a}$$

Your task is to develop a similar solution to the general system of two linear equations in two variables.

Such a system is shown here. The coefficients and constant terms are represented by the variables a, b, c, d, e, and f. Treat the variables x and y as the "unknowns."

$$ax + by = c$$
$$dx + ey = f$$

1. Find a general solution to this system of linear equations. That is, solve for x and y in terms of a, b, c, d, e, and f.

2. Discuss whether your solution makes sense for all values of the variables a, b, c, d, e, and f. Explain how this is related to the concepts of **inconsistent** and **dependent systems.**

Playing Forever

1. The high school baseball team has seven excellent outfielders: Andrew, Brett, Carl, Duane, Ethan, Felipe, and Griswold. Only three can start a given game.

   What is the most games the team can play before they have to repeat the same starting group of three outfielders? It doesn't matter which player is at which outfield position—simply consider which three are chosen and list all possible combinations.

2. The basketball team has nine excellent players: Hanna, Isabella, Jacinda, Katherine, Lana, Madeleine, Nikki, Oriana, and Paulette. Only five can start a given game.

   What is the most games the team can play before they have to repeat the same starting group? List all possible combinations.

The Shortest Game

In the game SubDivvy (from the POW *SubDivvy*), there are often many ways to get from the starting number to 1.

For example, if the starting number is 10, the game can proceed in several ways, including the two shown below. The first path takes four steps, while the second takes five steps.

$$
\begin{array}{r}
10 \\
-5 \\
\hline
5 \\
-1 \\
\hline
4 \\
-2 \\
\hline
2 \\
-1 \\
\hline
1
\end{array}
\qquad
\begin{array}{r}
10 \\
-1 \\
\hline
9 \\
-3 \\
\hline
6 \\
-3 \\
\hline
3 \\
-1 \\
\hline
2 \\
-1 \\
\hline
1
\end{array}
$$

Examine whether the shortest path to 1 is always achieved by subtracting the greatest possible divisor at each step. Then see what else you can find out about shortest paths in SubDivvy.

Producing Programming Problems— More Variables

In the Year 2 unit *Cookies,* you invented your own linear programming problem involving two variables. Your task now is to invent a linear programming problem involving four or more variables.

Here are the key items your problem must have.

- Four or more variables
- Something to be maximized or minimized that is a linear function of those variables
- Some linear constraints involving your variables (Some of these constraints can be equations instead of inequalities.)

Write your problem, and then solve it.

Your Own Three-Variable Problem

In the activity *Ages, Coins, and Fund-Raising,* you examined three problems. Each of them could be represented by a system of three linear equations in three variables.

Make up a problem of your own that can be represented that way, and solve your problem.

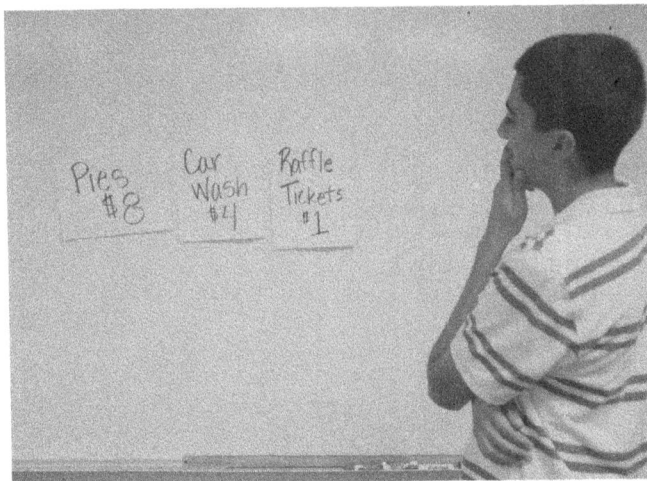

Fitting a Plane

In the activities *Fitting a Line* and *Fitting More Lines,* your task was to find a linear function through a particular pair of points. That is, you were asked to find coefficients a and b so that the line $y = ax + b$ would go through the given points.

This activity adds another dimension to this task—literally. Instead of finding the equation of a line through two points, you're asked to find the equation of a plane through three points.

Consider functions in which z is written as a linear expression in terms of x and y. That is, consider functions defined by equations of the form $z = ax + by + c$, where a, b, and c are any three numbers. The graph of any such equation is a plane.

Comment: A linear equation involving only x and y, such as $2x - 3y = 7$, can be graphed in 3-space, and its graph is a plane (just as an equation like $x = 3$ can be graphed in the xy-plane, and its graph is a line). Thus, the form $z = ax + by + c$ does not represent the most general plane.

1. Suppose you want a plane that goes through the point $(3, 2, 1)$.

   a. Show that the graph of the equation $z = 2x - 4y + 3$ goes through this point.

   b. Find a condition on a, b, and c that guarantees the graph of the equation $z = ax + by + c$ goes through $(3, 2, 1)$.

2. Find a condition on a, b, and c that guarantees the graph of the equation $z = ax + by + c$ goes through $(1, -2, 5)$.

3. Find a condition on a, b, and c that guarantees the graph of the equation $z = ax + by + c$ goes through $(2, -1, 6)$.

4. Find values of a, b, and c so that the graph of the equation $z = ax + by + c$ goes through all three points: $(3, 2, 1)$, $(1, -2, 5)$, and $(2, -1, 6)$.

5. Can you pick any three points and then find an equation of the form $z = ax + by + c$ whose graph goes through all three points?

Surfer's Shirts

1. Anna is ordering special T-shirts for the annual Cross Town Race. T-shirts come in two sizes: small and large. A small T-shirt costs $8 for the shirt itself and $2 to print the design. A large T-shirt costs $10, and it costs $2.50 to print the design. Anna needs to itemize the printing costs separately from the cost of the blank shirts.

   Suppose Anna orders S small shirts and L large shirts, with designs. Set up matrices with all of the necessary information. Write an expression using the matrices that will give Anna the total cost for the T-shirts and the total cost for printing (in terms of S and L).

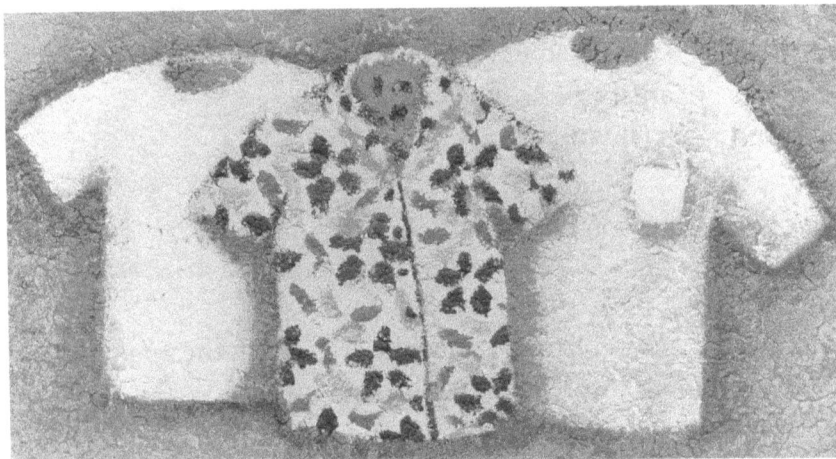

2. Ming's latest surfing competition is conducted in three heats. Ming is using her three classic moves: the off-the-lip, the cutback, and the tube ride.

   In this competition, off-the-lips are worth 4 points each, cutbacks are worth 6 points each, and tube rides are worth 10 points each.

   - In heat 1, Ming did 5 off-the-lips, 3 cutbacks, and 2 tube rides.
   - In heat 2, Ming did 7 off-the-lips, 2 cutbacks, and 1 tube ride.
   - In heat 3, Ming did 3 off-the-lips, 4 cutbacks, and 3 tube rides.

continued ▶

a. Suppose Ming wants to know how many points she scored in each heat. Set up matrices and write a matrix expression that will give her this information.

b. Suppose instead Ming wants to know the total number of points she scored (in the three heats combined) for each type of move. Set up matrices and write a matrix expression that will give her this information.

An Associative Proof

In the activity *Things We Take for Granted*, you investigated whether matrix multiplication is commutative or associative. This activity follows up on that work.

1. Prove that multiplication of 2 × 2 matrices is associative.

2. a. Is matrix addition commutative?

   b. Is matrix addition associative?

   Justify your answers.

When Can You Find an Inverse?

In the activity *Inverses and Equations,* you saw that some 2×2 matrices have an inverse for multiplication and others do not. You will now investigate the existence of matrix inverses more fully.

1. Think about the connection between whether a matrix has an inverse and the graphs of certain equations related to the matrix. Use the geometry of these graphs to find out which 2×2 matrices have inverses. Explain your answer.

2. Which of these 3×3 matrices have inverses? Explain your answers.

a. $\begin{bmatrix} 1 & 2 & 1 \\ 0 & 2 & -1 \\ 2 & 4 & 2 \end{bmatrix}$

b. $\begin{bmatrix} 1 & -1 & 2 \\ 0 & 3 & 0 \\ 3 & -3 & 1 \end{bmatrix}$

c. $\begin{bmatrix} 1 & -1 & 2 \\ 0 & 3 & 0 \\ 1 & 2 & 2 \end{bmatrix}$

d. $\begin{bmatrix} 1 & 2 & 1 \\ 0 & 2 & -1 \\ 2 & 6 & 2 \end{bmatrix}$

3. Based on your work in Question 2 and other examples of your own, what can you say about when a 3×3 matrix has an inverse and when it does not?

Determining the Determinant

If a system of linear equations has the same number of equations as variables, it will usually have a unique solution. In some cases, the system will have no solution (an inconsistent system) or infinitely many solutions (a dependent system).

Whether a system has a unique solution depends on the coefficients of the variables. As you may have seen in the activity *The General Two-Variable System,* the system

$$ax + by = c$$
$$dx + ey = f$$

will have a unique solution whenever a certain expression involving a, b, d, and e is not zero. When that expression is zero, the system will be either dependent or inconsistent, depending on the values of c and f.

This special expression involving the coefficients is called the *determinant* of the matrix $\begin{bmatrix} a & b \\ d & e \end{bmatrix}$.

In fact, determinants are defined for square matrices of all sizes and can be used to solve systems of linearequations.

Your Task

Your task is to learn more about determinants and to write a report about what you learn.

Begin your report with the definition of the determinant for 2 × 2 matrices. Explain how determinants can be used to solve two-variable systems of equations like the one shown in this activity.

Then discuss how determinants are used in general to solve systems of linear equations and what the relationship is between the determinant of a matrix and the invertibility of the matrix. Go beyond the 2 × 2 case and include the definition of the determinant for at least the case of a 3 × 3 matrix.

$$
\begin{array}{r}
A\ B \\
\times\ \ \ C \\
\hline
D\ E
\end{array}
\qquad
\begin{array}{r}
D\ E \\
+\ F\ G \\
\hline
H\ I
\end{array}
$$

Cracking Another Code

These two arithmetic problems use a single code in which letters have been substituted for numbers, as in the POW *Crack the Code*.

As in that POW, any solution must follow these rules.

- If a letter is used more than once, it stands for the same number each time it is used.
- Different letters always stand for different single-digit numbers.
- A letter standing for 0 never starts a number with more than one digit. For example, the expression 03 cannot be used, but 507 and 80 and just plain 0 are all allowed.

Your job is to crack the code. Keep in mind that the multiplication problem and the addition problem use the same code.

You must not only find all the solutions, if there are any, but also prove your answer.

Adapted from a problem in *Oregon Mathematics Teacher,* January–February 1989.

PHOTOGRAPHIC CREDITS

Front Cover Photos

(From top left, clockwise) Jonathan Wong, Johnny Tran, Dylan Matthews, Thao Nguyen, Eden Ogbai

Front Cover and Unit Opener Photography

Berkeley High School and Lincoln High School: Stephen Loewinsohn
Stock photos: iStockphoto

Meadows or Malls?

3 Lincoln High School, Stephen Loewinsohn; **4** iStockPhoto; **7** Rayman/ Photolibrary; **11** Lincoln High School, Stephen Loewinsohn; **13** Lincoln High School, Stephen Loewinsohn; **19** nick baylis/Alamy; **23** Lincoln High School, Stephen Loewinsohn; **24** Photodisc; **32** Hillary Turner; **33** Hillary Turner (both); **38** Lincoln High School, Stephen Loewinsohn; **40** Lincoln High School, Stephen Loewinsohn; **41** Shutterstock; **42** iStockPhoto; **43** Shasta High School, Dave Robathan; **44** iStockPhoto; Shutterstock; Hillary Turner; **45** Hillary Turner; **46** iStockPhoto; Hillary Turner; **52** Lincoln High School, Stephen Loewinsohn; **62** David Forbert/ Superstock; **63** Thinkstock/Getty Images; **66** iStockPhoto; **67** Lincoln High School, Stephen Loewinsohn; **68** Lincoln High School, Stephen Loewinsohn; **70** Hillary Turner (all); **73** Shutterstock; **76** Hillary Turner; **78** iStockPhoto; **80** Lincoln High School, Stephen Loewinsohn; **81** Lincoln High School, Stephen Loewinsohn; **82** Redlink Production/Corbis; **89** Lincoln High School, Stephen Loewinsohn; **91** Santa Maria High School, Chris Paulus; **93** iStockPhoto & Rayman/Photolibrary; **95** Hillary Turner; **96** Hillary Turner; **97** Lincoln High School, Stephen Loewinsohn; **105** Shutterstock; **106** Photodisc; **109** Shutterstock; **110** Lincoln High School, Stephen Loewinsohn; **112** Lincoln High School, Stephen Loewinsohn; **113** Shutterstock

www.ingramcontent.com/pod-product-compliance
Lightning Source LLC
Chambersburg PA
CBHW080426220326
41519CB00071BA/7217